Lecture Notes in Economics and Mathematical Systems

398

Founding Editors:

M. Beckmann
H. P. Künzi

AF856321

Managing Editors:

Prof. Dr. G. Fandel
Fachbereich Wirtschaftswissenschaften
Fernuniversität Hagen
Feithstr. 140/AVZ II, W-5800 Hagen 1, FRG

Prof. Dr. W. Trockel
Institut für Mathematische Wirtschaftsforschung (IMW)
Universität Bielefeld
Universitätsstr. 25, W-4800 Bielefeld 1, FRG

Jean-Pierre Urbain

Exogeneity in Error Correction Models

Springer-Verlag
Berlin Heidelberg New York
London Paris Tokyo
Hong Kong Barcelona
Budapest

Author

Dr. Jean-Pierre Urbain
University of Limburg
Department of Quantitative Economics
P.O. Box 616
NL-6200 MD Maastricht, The Netherlands

ISBN-13: 978-3-540-56639-7 e-ISBN-13: 978-3-642-95706-2
DOI: 10.1007/978-3-642-95706-2

Library of Congress Cataloging-in-Publication Data.

Urbain, Jean-Pierre.
Exogeneity in error correction models/Jean-Pierre Urbain. p. cm. - Lecture notes in economics and mathematical systems; 398)
Includes bibliographical references and index.
ISBN-13: 978-3-540-56639-7
1. Error analysis (Mathematics) 2. Economics-Statistical methods. I. Title. II. Series.
HB 137.U72 1994 330'.01'5195-dc20

Typesetting: Camera ready by author
42/3140-543210 - Printed on acid-free paper

Acknowledgements

The material contained in this monograph results of research realised when I was working at the University of Liège. On looking back at the process of writing this monograph, I realise that many people have contributed to its realisation.

First of all, I want to express my great indebtedness to Herman K. van Dijk, for having guided my work over the recent years. His constant support, both morally and intellectually, has been a major motivation for me to pursue this research. He has always been willing to read, comment and discuss the topics on which I have been working. Without his encouragments and suggestions, this work would certainly never have been conducted.

Over the recent years, I have also benefited from numerous discussions with Jim Lindsey. He has always been an astute and friendly critic who helped me clarify the arguments at various points. His careful readings of different versions of the manuscript, and more generally of my work, has led to substantial improvement and clarification in the material contained in this book. Luc Bauwens and Franz Palm contributed many important comments and suggestions for which I am extremely grateful. I also would like to express my gratitude to Yvan Langaskens for his support.

Several parts of this monograph have benefited from (oral or written) comments from Phillip Hans Franses, David Hendry, Sören Johansen and Mark Steel. In particular, I would like to thank Peter Boswijk for many stimulating discussions and for collaboration in work on exogeneity tests. Thanks are also due to Albert Corhay, Bernard Delhausse, Axel Lutgens and Sergio Perelman for discussions, suggestions and encouragements.

Financial support of the Faculty of Economics and Business Administration of the University of Limburg (Post-doctoral fellowship 1992-93) is gratefully aknowledged.

Finally, I want to acknowledge the support and help of my parents, friends and especially of my wife and my two daughters.

These acknowledgements would not be complete if I did not thank all those whose music has accompanied me in my late hours of work.

This monograph is dedicated to Marina, Emilie and Céline who had to live through the pressure during the realisation of this work. I want to thank them for their love.

Contents

List of Figures ix

List of Tables xi

1 Introduction and Summary 1

2 Cointegrated Systems 7

2.1 Some Historical Background to the Modelling of Economic Time Series 9

2.2 Integration and Cointegration 17

2.3 The Modelling of Cointegrated Systems 24

2.4 Cointegration and Conditional Sub-systems 34

2.5 Error Correction Models 38

2.6 Conclusions 40

3 Weak Exogeneity in ECMs 43

3.1 Weak Exogeneity 44

3.1.1 Definition and example 45

3.1.2 Empirical motivations 50

3.2 Reduced Form Error Correction Models 51

3.2.1 The error correction system in reduced form 51

3.2.2 Single equation error correction model in reduced form and weak exogeneity 53

3.3 ECMs in Structural Form 61

3.3.1 The error correction system in structural form 61

3.3.2 Single equation error correction model in structural form and weak exogeneity 62

3.4 Inference on Weak Exogeneity in ECMs 65

3.4.1 Are orthogonality tests useful in single equation ECMs 65

3.4.2 Testing for the presence of cointegrating vectors in the marginal models 67
3.5 Empirical Illustration 68
3.5.1 Belgium consumption function: Steel (1987) 68
3.5.2 Aggregate import demand function: Urbain (1988) . . 72
3.5.3 The UK demand for money: Lubrano et al. (1986), Steel and Richard (1991) 73
3.6 Conclusions 81

4 Testing for Weak Exogeneity 83
4.1 Introduction 83
4.2 Exogeneity and the Incomplete SEM 85
4.3 The Behaviour of Orthogonality Tests in the Presence of (Co)-Integrated Variables 90
4.3.1 Testing for weak exogeneity within a limited information framework 91
4.3.2 Small sample behaviour: some simulation evidence . . 93
4.3.3 Asymptotic distribution 101
4.4 Testing for Weak Exogeneity in ECMs where the Short Run Dynamic Parameters are Parameters of Interest 107
4.5 Conclusions 111

5 Empirical Analysis: The Case of Aggregate Imports 113
5.1 Background 113
5.1.1 The imperfect substitutes model for aggregate trade flows 114
5.1.2 Econometric issues in trade modelling 116
5.2 System versus Partial Approach to the Modelling of Belgium Aggregate Imports 123
5.2.1 Univariate analysis 124
5.2.2 Multivariate cointegration and simultaneous equation models 137
5.2.3 Partial modelling of aggregate imports: a structural ECM 149
5.3 Conclusions 160

6 Conclusions 161

Bibliography 165

Author Index 185

Subject Index 188

List of Figures

2.1 Stationary process 20
2.2 $I(1)$ process 20
2.3 near-$I(1)$ process 21
2.4 Independent $I(1)$ processes 23
2.5 Bivariate cointegrated series 23

3.1 Real disposable income and M3 (in logs) 75
3.2 Deflator of the real disposable income and short term interest rates (in logs) 75
3.3 Deflator of the real disposable income (in first diff.) 76
3.4 Short run interest rates (in first diff.) 76
3.5 M3 (in first diff.) 77
3.6 Real personal disposable income (in first diff.) 77

5.1 Import and domestic prices (in level) 125
5.2 Import prices (in first diff.) 125
5.3 Domestic prices (in first diff.) 126
5.4 Import volume (in level) 126
5.5 Import volume (in first diff.) 127
5.6 Real income (in level) 127
5.7 Real income (in first diff.) 128
5.8 Sequential ADF tests: import volume 135
5.9 Sequential ADF tests: real income 136
5.10 Sequential ADF tests: domestic prices 136
5.11 Sequential ADF tests: import prices 137
5.12 Restricted cointegrating vector 143
5.13 1-step Chow tests (PVAR): import volume equation 145
5.14 1-step Chow tests (PVAR): real income equation 146
5.15 1-step Chow tests (PVAR): domestic prices equation 146

5.16 1-step Chow tests (PVAR): import prices equation 147
5.17 N-step ahead Chow tests: structural ECM 158
5.18 N-step backward Chow tests: structural ECM 159
5.19 1-step Chow tests: structural ECM 159

List of Tables

3.1 Auxiliary regression for Δia_t^* 70
3.2 Exogeneity tests by variable addition 71
3.3 Money demand in the UK: MLE approach with $n = 3$ 79
3.4 Money demand in the UK: MLE approach with $n = 5$ 80

4.1 Representation of incomplete SEMs 88
4.2 Instruments used in the simulations 96
4.3 Empirical size of orthogonality tests 98
4.4 Empirical power of orthogonality tests , $\rho = 0.25$ 99
4.5 Empirical power of orthogonality tests , $\rho = 0.50$ 100

5.1 Testing for seasonal unit roots 130
5.2 Testing for unit roots at the zero frequency 132
5.3 Testing for breaking trend: the 1973 oil shock 134
5.4 Lag length selection and Johansen's cointegration tests 139
5.5 Multivariate cointegration analysis 141
5.6 Hypothesis tests on cointegrating matrix 142
5.7 Statistics on the error processes 144
5.8 FIML estimation of the final SEM 148
5.9 Wald tests for Cointegration 151
5.10 Cointegrating vector estimates 152
5.11 Structural ECM and weak exogeneity 155
5.12 Weak exogeneity tests 156
5.13 ECM for aggregate imports 157

Chapter 1

Introduction and Summary

The last two decades have witnessed a number of important methodological contributions to the empirical modelling of economic time series. Traditional econometric practice has been widely questioned both for its statistical fragility and its questionable economic implications (see the papers collected in Granger, 1990). In particular, the antagonism between pure time series analysis in the Box-Jenkins (1970) tradition and the so called "textbook" approach to econometric (for example Johnston, 1984) has disappeared in favor of empirical modelling approaches taking fully advantage of time series analysis for the elaboration of well behaved dynamic econometric models.[1]

This new wave of research has resulted in the elaboration of new concepts, methods and approaches to the dynamic empirical modelling of economic time series (see the overview in Gourieroux, 1990). Among these, the most significant breakthrough in econometric practice is the explicit recognition of the (possible) nonstationary character of most macroeconomic time series, the latter usually well represented by stochastic processes accumulating shocks over time and therefore labelled integrated processes. The implication in terms of deviation from usual asymptotic theory (see inter alia Park and Phillips, 1988, 1989) has led to the emergence of a new wave of research concerned with the efficient estimation of stable relationships among nonstationary variables. This resulted in the notion of cointegration introduced by

[1]A brief account of these developments is presented in the beginning of the second chapter.

Granger (1981), Granger and Weiss (1983) and especially popular since its presentation by Rob Engle and Clive Granger at the Fifth World Meeting of the Econometric Society in 1985. Since then, the study of cointegrated time series, i.e. nonstationary series which move stochastically together over time, has become one of the more prolific and promising fields of research in theoretical and applied econometrics and also in statistical analysis of time series. However, numerous different approaches have been advocated for the efficient estimation of, and testing for, cointegrating relationships among nonstationary economic variables. In a recent overview, Phillips and Loretan (1991) investigate a dozen different approaches for the estimation of cointegration relationships and study their relative merits in terms of asymptotic efficiency. While such a paper has the great benefit of clearly stating problems in terms of asymptotic efficiency of cointegration vector estimators, it nevertheless leaves the applied researcher confronted with a wide variety of possible alternative approaches without providing a real guide for empirical research. In this respect, it is worth noting several Monte Carlo studies of the finite sample behaviour of alternative cointegration tests and cointegration estimators (Gonzalo, 1989, Gregory, 1991, Reimers, 1991, Park and Ogaki, 1991, Boswijk, 1992a,b) and especially the recent surveys of Campbell and Perron (1991, 1992) who propose rules the applied researcher has to be aware of when studying relations among nonstationary macroeconomic times series. These rules cover inference on the presence of unit roots in univariate times series as well as in multivariate time series. They insist on two major points:

> *... the importance of correctly specifying deterministic components of the series; and the second, the usefulness of unit root tests not as methods to uncover "true relation" but as practical devices that can be used to impose reasonable restrictions on the data and to suggest what asymptotic distribution theory gives the best approximation to the finite sample distribution of the coefficient estimates and test statistics.*

However, even in this extensive review of alternative estimators, tests and model representations, there remain at least two unanswered questions: (*i*) knowing whether we should conduct the analysis by explicitly modelling the short run dynamic or not, (*ii*) whether cointegrated processes must be investigated in closed multivariate systems, in partial systems (where the generating model of some variables is not modelled) or in single equation framework. This last question is partly addressed in the recent survey of Johansen (1991b) who considered the problems of the statistical modelling of systems of trending variables. Although the superiority of full systems based approaches has been clearly demonstrated among others by Phillips (1991a), conditional models have the same properties when exogeneity conditions are verified. In this respect, the question is very similar to the traditional one of Full versus Limited Information analysis of simultaneous equation models. If exogeneity holds, lower dimension systems are obviously, from a practical point of view, easier to handle. Moreover, in that case standard limiting distribution results apply to the cointegrating vector estimates (see Johansen, 1992a, Phillips, 1991a, Boswijk, 1991a, 1992a,b).

The empirical researcher is thus often confronted with the choice of running the risk of losing efficiency or losing empirical tractability. As exogeneity is essential for the efficiency of conditional models, the question that naturally emerges, is whether cointegrated systems have particular characteristics which make the investigation of exogeneity substantially different from the traditional stationary framework. This will be the major concern of this work where we discuss the concept of exogeneity in cointegrated systems.

In particular, we try to point out the potential advantages and drawbacks of working in a conditional framework or in complete models. We do not investigate the asymptotic behaviour of cointegration tests or the asymptotic efficiency of competitive estimators since important contributions have been provided recently (see Phillips, 1991a, Phillips and Loretan, 1991). Instead, we focus on a particular representation of cointegrated systems, namely Error Correction Models (ECMs), which have the important advantage of explicitly modelling the short run dynamic jointly with the long run relationships. We basically think that from an economic point of view these short run parameters have important implications for forecasting, policy simulations and interpretation. However, they have sometimes been neglected in recent theoretical contributions to the study of cointegrated processes, often by non-parametrically correcting for their effects. Exceptions to this trend are the contributions of some authors (Johansen, 1988, Dolado et al., 1989, Kremers et al. 1992, 1991, Boswijk, 1990, 1991a, 1992a,b) who thereby extend the more traditional Hendry-Richard (1982, 1983) -see also Hendry (1987)- approach to dynamic empirical econometric modelling.

The outline of the work is as follows. In Chapter 2, some of the major developments in the study and modelling of economic time series, which have led to this new wave of research, are mentioned. We focus on concepts and basic models so that the presentation is, as far as possible, a nontechnical overview of the fundamentals underlying our work. Some alternative ways of modelling cointegrated systems are presented with a special emphasis given to the approaches that are followed in the empirical part of this work. We do not give an exhaustive overview of the subject, the interested reader is referred to recent surveys on the topic for more extensive presentations. See inter alia Banerjee and Hendry (1992), Campbell and Perron (1991, 1992), Dolado et al. (1990), Diebold and Nerlove (1990), Engle and Yoo (1991), Johansen (1991b), Phillips and Loretan (1991), Boswijk (1992b) as well as the special issues of the Oxford Bulletin of Economics and Statistics (1986, 1992), that of the Journal of Economic Dynamic and Control (1988) and the recent book edited by Engle and Granger (1991).

In Chapter 3, we draw upon the results reported in Urbain (1992b) and investigate the concept of weak exogeneity in Error Correction Models (ECMs). More particularly we consider error correction models such as those that have been widely applied within the Hendry approach. Since derived models do not always have parameters which are easily interpretable in terms of economic notions, we also consider models where some structure has been imposed by allowing for (over)identification restrictions. In the light of the implications of cointegration on the factorisation of a vector autoregressive model, we derive the conditions of weak exogeneity and their consequences for inference. It is found that the particular form of error correction model which is used implies different conditions of exogeneity and that, in almost all cases, the usual orthogonality condition is not sufficient for weak exogeneity.

In Chapter 4 we consider exogeneity tests in error correction models, the underlying model representation and the potential defects and usefulness of traditional orthogonality tests. This is first investigated by means of a small illustrative Monte Carlo experiment. The asymptotic behaviour of the Hausman (1978) tests is also derived. Finally we propose a simple variable addition procedure for testing weak exogeneity of an explanatory variable with respect to both the short and long run dynamic parameters in structural error correction models.

Chapter 5 is an empirical study of aggregate imports for a small and open economy: Belgium. We first review the usual way aggregate imports are modelled and point out some important pitfalls of this. We then consider two alternative approaches, one based on the development of a complete simultaneous model following the recent work of Hendry and Mizon (1990), while the second approach is due to Boswijk (1990, 1991a) and consists of error correction models in structural form. In both cases, the way exogeneity can be analysed is empirically implemented.[2]

It must be clearly stated that our empirical work is conducted in the modelling framework which has been advocated by Hendry and Richard (1983, 1982), Hendry (1987), Spanos (1988), Hendry and Mizon (1990). We will therefore not discuss the various controversies that exist on the optimal way empirical analysis should be conducted. These are extensively presented in the recent discussion reported in Poirier (1990), and also summarised in Granger (1990) which presents most of the modelling concepts we are using. A model will only be considered as an approximation to the true Data Generating Process (DGP) which is unknown, unless a Monte Carlo experiment is conducted. It is clear therefore that several models might be equally well approximating this unknown data generating process and that the use of adequate model design criteria will hopefully allow us to derive reasonably well behaved approximations to the DGP. The criticism of data mining is naturally applicable, but we think that looking, even too much, at the data is preferable to looking too much at theory, although the optimal strategy certainly lies somewhere in between these two extremes.

A summary of the major findings of this monograph, general conclusions and some routes for further research are reported in a concluding chapter.

[2]Note that we work in a classical framework. Developments in Bayesian analysis of the simultaneous equation model and time series models will not be considered.

Chapter 2

Cointegrated Systems

In this introductory chapter, we present and discuss some alternative approaches to the modelling of cointegrated systems. In particular, we try to put forward arguments for and against the modelling of cointegration relationships based on closed models or on conditional sub-systems. The structure of the chapter is as follows. We first clarify the basic notation that is used in this chapter. As the focus on the nonstationary properties of economic time series is relatively recent in econometrics, Section 1 reviews some developments of macroeconometric model building which have led to the notion of cointegration. Section 2 discusses the modelling of cointegrated series and particularly closed system (parametric) approaches. In Section 3, we discuss the use of (open) conditional systems while a brief discussion of the interpretation of ECMs is presented in Section 4.

Notation

Let us first clarify the notation used in this study.

- $\{x_t\}$: is a $k \times 1$ vector time series, for which we have a sample of T observations, e.g. x_t is the column-vector containing the tth observations on a set of k variables.

- X_{t-1} is the information set available at time $t-1$ defined by

$$X_{t-1} = (X_0, x_1, \ldots, x_{t-1})'$$

where X_0 is a matrix of initial conditions.

- When needed, the vector x_t is partitioned into two subgroups of variables $(y_t, z_t)'$ which respectively denote variables which we want to model explicitly (endogenous variables) and those whose generating model will not be modelled and thus on which we shall condition ("exogenous" variables).

- L is defined as the lag operator L, such that, for a time series y_t $(t = 1, 2, \ldots)$, $Ly_t = y_{t-1}$. More generally, the lag operator may be raised to any (integer) power such that

$$L^k y_t = y_{t-k}$$

The extensive use of expressions in L allows for important notational simplifications in a time series context. Notice that the algebra of the linear backshift operator (B), as described for example in Dhrymes (1981), is applicable to the lag operator L. In the sequel we make extensive use of polynomials in L, i.e. *lag polynomials*:

$$\Pi(L) = \Pi_0 + \Pi_1 L + \Pi_2 L^2 + \ldots = \sum_{i=0}^{\infty} \Pi_i L^i$$

For a given polynomial $\Pi(L)$, we shall also use $\Pi(1)$, obtained by replacing the argument L by 1 in $\Pi(L)$:

$$\Pi(1) = \Pi_0 + \Pi_1 + \Pi_2 + \ldots = \sum_{i=0}^{\infty} \Pi_i$$

More generally, it must be noted that the algebra of polynomial operators in L is equivalent to the usual algebra of polynomial functions. Similarly, it is straightforward to extend this notation to the case of a *matrix polynomial* in L, in which case the element Π_i of a given polynomial $\Pi(L)$ becomes a square matrix. For example, if we consider a $k \times 1$ vector time series x_t, the matrix polynomial $\Pi(L)x_t$ can be written in detail as $\sum_{i=0}^{\infty} \Pi_i L^i x_t$ where the Π_i's are $k \times k$ matrices. We shall also use the difference operator Δ defined as $\Delta = 1 - L$.

- For simplicity, and also tractability, we limit ourselves to linear models and to the Gaussian distribution which is maintained as sufficiently well suited for the problems at hand.

2.1 Some Historical Background to the Modelling of Economic Time Series

The last two decades have witnessed a number of important methodological contributions to the empirical modelling of economic time series. In particular, the antagonism between pure time series analysis and econometrics has disappeared in favor of empirical modelling approaches taking fully advantage of time series analysis for the elaboration of well behaved dynamic econometric models. In the particular field of macroeconomics, a major modification in the practice of modelling economic series has been introduced with the recognition of the nonstationary character of most economic time series. It seems therefore interesting briefly to review some developments which have led to this new way of investigating economic time series. Without pretending to be exhaustive or complete we mention some major steps which, we think, have led to the emergence of new concepts and methods.

Since the seminal work done at the Cowles Commision (Koopmans, 1950, Koopmans and Hood, 1953),[1] it has been traditional to model jointly economic relationships among different economic variables through Simultaneous Equation Models (SEMs) of the form:

$$By_t + \Gamma z_t = u_t \tag{2.1}$$

where $u_t \sim N(0, \Sigma)$ with Σ a non-diagonal positive definite symmetric covariance matrix. The matrices B and Γ are composed of unknown parameters and known values (such as zeros and ones) which arise from economic theory. However, the lack of empirical evidence in favor of these type of models, and the very frequent rejection of the economically essential hypothesis related to B and Γ, has also (see the discussion in Anderson, 1991) resulted in the generalisation to dynamic simultaneous equation models (Koopmans, Rubin and Leipnik, 1950) where the matrices B and Γ are replaced by matrix

[1]See also Epstein, 1987, Malinvaud, 1988, Anderson, 1991

polynomials in the lag operator $B(L)$ and $\Gamma(L)$ with $B(L) = \sum_{i=0}^{k} B_i L^i$ and $\Gamma(L) = \sum_{i=0}^{p} \Gamma_i L^i$:

$$B(L)y_t + \Gamma(L)z_t = u_t \tag{2.2}$$

By allowing the "disturbances" u_t to be generated by general ARMA processes, these models were a potential candidate for the presentation of many macroeconomic phenomena (Zellner and Palm, 1974).

In the mid-seventies however, the occurrence of the first oil shock resulted in the breakdown of most forecasts based on traditional simultaneous macroeconometric models, so that time series models competed well, in terms of forecasts, against large SEMs. Important methodological contributions were made which substantially modified the way macroeconomic time series should be modelled by providing arguments in favor of the extensive use of time series analysis for the specification and elaboration of dynamic econometric models. For example, the link between SEMs and pure time series models has been precisely emphasized by Zellner and Palm (1974) who showed that ARIMA models for the endogenous variables could be derived from traditional SEMs when the exogenous variables are generated by ARIMA processes. The SEM then appears as a special case of the multivariate ARIMA model[2]. This provided an interesting way of assessing the validity of the traditional dynamic SEM since dynamic misspecification analysis of the latter can be based on detailed statistical and forecasting analysis of the former. As pointed out in Palm (1990), this has become a standard practice and simple time series models are often used as a yardstick against which the forecasting performances of econometric models are tested (see Hendry, 1983, Davidson et al., 1978).

Multivariate time series models have also become very popular in econometrics due to the seminal paper by Sims (1980), reactualizing Liu's (1960) work. Sims (1980) criticised the foundation and even the validity of most a priori theoretic restrictions imposed in the framework of SEMs. To quote Sims (1980), *"these structural identifying restrictions are often too incredible to be able to yield econometric models which might be useful in any sense of the word"*. He therefore advises the use of **V**ector **A**uto**R**egressive (VAR) models as a simple class of models which can describe and capture the complex inter-variable dynamic relationships and generate a wide class

[2] For a good presentation of time series analysis we refer to Granger and Newbold (1986), Fuller (1976), Harvey (1981) Gourieroux and Monfort (1990).

of dynamic behaviour among a given set of variables without imposing identifying restrictions. The VAR(p) model for the x_t process is given by:

$$A(L)x_t = u_t \tag{2.3}$$

with

$$A(L) = I + A_1L + A_2L^2 + \ldots + A_pL^p$$

and u_t is assumed a white noise vector process, i.e. a process which is not predictable linearly from its past, beyond its unconditional mean which is taken to be zero. More formally, $\{u_t\}$ is a (mean) white noise process if $E(u_t|U_{t-1}) = E(u_t) = 0$.

As an alternative to dynamic SEMs, VAR models are characterised by the joint and closed modelling of all variables conditional on the entire past information set available at time $t-1$ so that no variables are incorrectly considered as exogenous. VAR models suffer from several defaults, a major one being that the number of parameters to estimate grows at the rate of k^2 for each additional lag. However they have become a standard tool for both forecasting and policy simulation through impulse response analysis. Moreover, Bayesian priors have been introduced in the model in order to shrink some coefficients towards zero resulting in more parsimonious VAR models (see Todd, 1990, for an overview of the developments of Bayesian VAR's and the recent textbook of Lütkepohl,1991, for a general treatment of VAR models).

A different point of view was adopted by several authors who perceived a priori economic theories as too important and too essential to be doubted on the sole basis of empirical stylised facts. Among these we must mention the well-known and often quoted *Lucas critique* (Lucas, 1976) which casts doubt on the usefulness of many models due to their inability to reflect the behaviour of economic agents in the face of a changing environment. Lucas (1976) therefore advised developing (theoretical based) structural models in terms of (to quote Sargent, 1981) "deep structural parameters" so that they should be able to reflect the principle that the agents' behaviours are influenced by their constraints. This approach has been further linked to time series analysis in the SEMSTA approach, developed and advocated by Palm (1983, 1990), which combines theoretical based dynamic stochastic models with extensive diagnostic and specification tests.

Somewhere between the VAR and the theoretical based approach, David Hendry and many of his co-authors have proposed an empirical based modelling approach for dynamic empirical econometric models (see Davidson et al., 1978, Hendry and Richard, 1982, 1983, Hendry, 1983). Although the "Hendry" approach has been widely discussed and presented in the recent literature (e.g. Pagan, 1987, Hendry, 1987, Ericsson et al., 1991, Cuthbersom et al., 1992) it may be useful to present, at least conceptually, the general essence of this modelling strategy which we follow implicitly in the empirical part of this work. The central concept is the **D**ata **G**enerating **P**rocess (DGP) which is a general statement of the joint probability distribution of all observable variables. The observed data are assumed to be generated by the unknown effect of the agents acting in the economic world. Empirical econometric models are then considered as highly simplified representations of it, i.e. they are seen as **reductions** of the unknown DGP. For practical purposes, one must still make some statements on the form of the DGP. We first assume that the DGP is represented by the joint probability distribution of all the observable variables:

$$D(w_1, \ldots, w_T | W_0; \theta) \quad \theta \in \Theta$$

where $D(.|.)$ denotes the density function for $\{w_t\}$, W_0 are initial conditions, θ is a vector of identifiable parameters of the distribution, T is the number of observations available in the sample and Θ is the parameter space, i.e. the variation domain of θ. As such, the DGP is clearly too general to be useful, being intrinsically complicated and involving, in w_t, a high number of potential economic variables. To be useful, and to enable the investigator to approximate this DGP, the latter has to be reduced by means of two types of *reductions*: **marginalisation** and **conditioning**.

In particular, one first (usually implicitly) *marginalises* with respect to all variables, and their past values, which are deemed to be irrelevant for the purpose of the study. At this stage, the use of economic theory, or empirical stylised facts, is often the implicit rule. Note that invalid "global" marginalisation can result in an important loss of information (see the discussion in Florens and Mouchart, 1985). The second step is to assume that economic activity is inherently sequential so that the joint density can be rewritten as

$$\prod_{t=1}^{T} D(x_t | X_{t-1}; \lambda) \tag{2.4}$$

where x_t is the reduced set of variables after global marginalisation (so that it is typically of lower dimension than w), $X_{t-1} = (x_{t-1}, \ldots, x_1, X_0)$ and the vector λ is the reparametrisation of θ arising from the global marginalisation and the sequential conditioning. Assuming a linear normal conditional representation of the mean of observables, we have

$$x_t | X_{t-1} \sim N(\mu_t, \Sigma_t) \tag{2.5}$$

with μ_t and Σ_t respectively the conditional mean and conditional variance of x_t.

Defining $\varepsilon_t = x_t - \mu_t$ results in "disturbances" $\{\varepsilon_t\}$ which are "derived" instead of simply added as in the traditional textbook approach to econometrics. By construction, the sequence $\{\varepsilon_t\}$ constitutes a martingale difference sequence (with respect to the filtration X_{t-1}), i.e. ε_t is a mean innovation process w.r.t. the X_{t-1}, so that it is possible to assess the validity of each subsequent reduction by checking the behaviour of the resulting residuals. Note that the (mean) innovation property is much stronger than that of white noise. Whether a given process is an innovation crucially depends on the information set considered. In particular, we say that the process $\{\varepsilon_t\}$ is a mean innovation process with respect to X_{t-1} if $E(\varepsilon_t | X_{t-1}) = 0, \forall t$.

In order to allow for empirical analysis, we must still add two assumptions on (2.5): we assume (i) time homogeneity and (ii) a finite fixed (usually low) $p-th$ order approximation of X_{t-1}. The conditional mean can then be written as

$$\mu_t = \sum_{i=1}^{p} \Pi_i x_{t-i} \tag{2.6}$$

which results in an approximation of the DGP which takes the form of a finite VAR(p) model:

$$x_t = \sum_{i=1}^{p} \Pi_i x_{t-i} + \varepsilon_t \tag{2.7}$$

where ε_t is, by construction, normally independently identically distributed (n.i.i.d.) with p.d.s. covariance matrix Σ and p is the lag truncation. This model serves as the yardstick against which the validity of each further reduction will be tested. The model (2.7) is the "unrestricted reduced form" in terms of traditional terminology, while Spanos (1988) refers to it as to the "statistical generating model" in contrast to the DGP and the theoretical (economic) model. Note that in such a general "derived" model, the parameters are often hardly interpretable or hardly linked to economically meaningful concepts. Moreover some parameters might show nonconstancy while functions of them do not. Thus, further reductions or transformations are desired to satisfy a set of criteria including, among others, parsimony, constancy.

In particular, the investigator might not be interested in modelling the whole set of variables present in x_t, so that he may want to **contemporaneously condition** on a subset of variables in x_t whose generating process is of no interest for the purpose of the study. This results in conditional models where a few variables are considered as "endogenous" and the others as "exogenous". Here too, invalid conditioning can have severe effects, both in terms of efficiency and parameter constancy. This is crucially related to the notion of **weak exogeneity** which we shall discuss in more detail in Chapter 3. At this stage, we may limit ourselves by mentioning that weak exogeneity means that the generating models of the variables on which we have conditioned do not contain essential information with respect to the parameters that the investigator wants to study. If the purpose of the model is also forecasting and/or policy simulation, then further conditions have to be satisfied (see Engle et al., 1983). In particular, when the purpose of the analysis is conditional forecasting, we have to add the condition that there is no feedback from the endogenous to the "weakly exogenous", i.e. Granger non-causality, resulting in the stronger concept of **strong exogeneity**. Finally, when policy simulations have to be performed from the conditional model, the latter has to be characterised by constant parameters even when the processes generating the exogenous variables are affected by important regime changes, i.e. the conditional model has to be invariant to changes affecting the generating processes of the exogenous variable. The latter condition, when added to that of weak exogeneity, results in the concept of **super exogeneity** whose invalid character is the essence of the Lucas critique. When the conditioning variables are super exogenous, then the model can be interpreted as a structural model. As pointed out in Ericsson (1992), super exogeneity effectively identifies the parameters in the sense of uniqueness. The structural form emerges thus as a reparametrisation of the conditional expectation relative to a given information set, i.e. as a reparametrisation of the "unrestricted reduced form" (see the discussion in Spanos, 1990a, Hendry and Mizon, 1990) contrary to the traditional textbook framework where the latter is deduced **from** the structural form.

In practice, most of the applications of the Hendry approach have been confined to a very small number of "endogenous" variables, usually one, so that the empirical analysis is a typical example of an incomplete simultaneous equation model (see Richard, 1979, 1984). It usually starts by fitting an unrestricted[3] Autoregressive Distributed Lag (ADL) model which arises by contemporaneous conditioning in a VAR model. For simplicity, we restrict our attention to the case of two variables[4] y_t and z_t, with $p = 1$:

$$y_t = \alpha + \beta_0 z_t + \beta_1 z_{t-1} + \beta_2 y_{t-1} + \varepsilon_t \tag{2.8}$$

with $\varepsilon_t \sim$ n.i.i.d.$(0, \sigma_\varepsilon^2)$.

Note that ADL models are very general as they cover an important number of more particular dynamic specifications as discussed in Hendry et al. (1984). Further reductions are then applied and the ADL model is carefully empirically reparametrised in a more parsimonious way, in an attempt to distinguish carefully between short run and long run effects. The structural interpretation is often related to the long run solution of the model although the specification used in the empirical analysis can also be theoretically justified as deriving from optimal agents' response in intertemporal contexts (Hendry and Von Urgen-Sternberg, 1981). An important characteristic of the empirical application of the "general to specific" modelling strategy is the use of several criteria (see Hendry, 1987, Hendry and Ericsson, 1991, Ericsson et al., 1991) an empirical model should meet in order to be assumed to approximate adequately the unknown underlying DGP. Each of these implies testable parametric hypotheses so that the modelling approach is viewed as an exercise in design. Refined concepts of exogeneity such as super exogeneity[5], when satisfied, guarantee that the empirical model is not subject to the Lucas critique.

A similar strategy for SEMs is advocated in Hendry, Neale and Sbra (1988) where the starting point is a dynamic unrestricted reduced form such as (2.7) against which the overidentifying restrictions of the SEM are then tested (see also Palm, 1983, Spanos, 1990).

[3] Except for the lag length which is assumed fixed

[4] Which we assume are in natural logarithm.

[5] For test statistics see Engle and Hendry (1990), and for an application see Lahaye and Urbain (1991).

The empirical success of the "Hendry" approach, and especially its extensive use of parsimonious Error Correction Models (ECMs), has without any doubt led to important theoretical developments in the analysis of economic time series. A prototypical ECM can simply be obtained from (2.8) by rearranging the terms:

$$\Delta y_t = \alpha + \beta_0 \Delta z_t + (\beta_2 - 1)(y_{t-1} - \delta z_{t-1}) + \varepsilon_t \tag{2.9}$$

with

$$\delta = \frac{\beta_0 + \beta_1}{1 - \beta_2}$$

provided that $\beta_2 \neq 0$ and $\beta_2 < 1$, the latter condition being required for dynamic stability.

ECMs generalise partial adjustment schemes without imposing restrictions on the dynamic responses. The change in y_t, Δy_t, is a linear function of the change occurring in z_t, Δz_t and of departures of $y_{t-1} - \delta z_{t-1}$, the latter term reflecting the idea that agents "correct" their past observed "errors".

If we now consider a *steady state growth* with $\varepsilon_t \equiv 0$, $\Delta y_t \equiv g_1$, $\Delta z_t \equiv g$, then from (2.8), and using the fact that $y_{t-1} = y_t - g_1$, $z_{t-1} = z_t - g$, we see that

$$g_1 = g \cdot \frac{\beta_0 + \beta_1}{1 - \beta_2}$$

Using this, (2.8) is shown to be equivalent to

$$y = \frac{\alpha}{1 - \beta_2} - g \cdot \frac{\beta_1 + \beta_2 \beta_0}{(1 - \beta_2)^2} + \frac{(\beta_0 + \beta_1)}{(1 - \beta_2)} z \tag{2.10}$$

so that δ in (2.9) corresponds to the long run elasticity of y with respect to z. Long run proportionality between y_t and z_t, i.e. $g = g_1$, is obtained under the restriction $\beta_0 + \beta_1 + \beta_2 = 1$. In this case, letting lowercase letters denote the natural logarithm of the corresponding variables, (2.10) implies

$$Y_t = K Z_t$$

where $K = \exp\{[\alpha - g(1 - \beta_0)]/(1 - \beta_2)\}$. Notice that the long run steady state solution contains the growth rate g unless $\beta_0 = 1$ or α is an appropriate function of g (see Kloek, 1984).

Similarly, if we assume a long run equilibrium situation with $y_t = y_{t-1} = y^*, z_t = z_{t-1} = z^*$ and $\varepsilon_t \equiv 0$ then (2.8) is equivalent to

$$y^* = \frac{\alpha}{1-\beta_2} + \delta x^* \tag{2.11}$$

Error correction models thus allow for the existence of steady state equilibrium (or long run equilibrium) relationships and adjusment costs where the agents correct their decision in light of their initial disequilibrium.

A further factor that has contributed to the growing popularity of ECMs is the introduction of the concepts of *integrated* and *cointegrated economic time series* which pointed out that ECMs emerge naturally when the time series behaviour of the variables is taken into account.

2.2 Integration and Cointegration

Although it was commonly accepted that most macroeconomic data series display strong pattern of secular growth over time, it was not really until the contribution of Nelson and Plosser (1982) that the *nature* of this observed nonstationarity was extensively studied and finally often recognised to be of "stochastic" nature. In contrast to the practice of considering the observed nonstationarity to be of deterministic nature, the series are often empirically well characterised by *difference stationary processes*, i.e. series which become stationary after suitable differencing.[6] The underlying processes are called **integrated processes** for they accumulate past random shocks, each of which has an enduring effect on the historical behaviour of the data series.

Since integrated processes are a specific class of nonstationary processes, we first define what is meant by stationarity (see for example Spanos, 1986, Gourieroux and Monfort, 1990):

Definition 1
The stochastic process $\{x_t\}_1^\infty$ *is said to be* **stationary** *for any admissible* $t_1, \ldots, t_n$ *and any* τ*, if the joint probability distribution of* $(x_{t_1}, \ldots, x_{t_n})$ *is identical to the joint probability distribution of* $(x_{t_{1+\tau}}, \ldots, x_{t_{n+\tau}})$

[6]Therefore labelled Difference Stationary in contrast to traditional Trend Stationary processes.

In particular we consider second order -or covariance- stationary processes so that the mean and variance are independent of time, while the autocovariance at lag τ only depends on τ. Following Granger (1981), we introduce the definition of an integrated process as follows:[7]

Definition 2
A time series x_t which has a stationary, invertible non-deterministic ARMA representation after differencing d times is **integrated of order d** *and is denoted by $x_t \sim I(d)$.*

The simplest example of an $I(1)$ process, i.e. integrated of order 1, is the random walk model with Gaussian innovations and nonzero drift μ:

$$x_t = \mu + x_{t-1} + \varepsilon_t \tag{2.12}$$

where for simplicity we can allow x_0 to be a known constant while ε_t is assumed to be n.i.i.d.$(0, \sigma^2)$. When allowing for a non-zero positive drift (i.e. $\mu \neq 0$), the series underlying x_t becomes a linear function of time with slope μ and with nonstationary deviation from this function which accumulates past random shocks giving rise to the notion of integrated series. This is obvious by rewriting the random walk as

$$x_t = \mu\, t + \sum_{j=1}^{t} \varepsilon_j + x_0 \tag{2.13}$$

Given that ε_t is assumed to be n.i.i.d.$(0, \sigma^2)$, x_t has the following first and second moments:

$$E(x_t|x_0) = x_0 + \mu t \tag{2.14}$$

$$\mathrm{Var}(x_t|x_0) = t\sigma^2 \tag{2.15}$$

$$\mathrm{Cov}(x_t, x_{t-\tau}|x_0) = (t-\tau)\sigma^2 \tag{2.16}$$

so that x_t is clearly nonstationary. Note directly that the asymptotic theory for stationary processes does not carry over to integrated processes since

[7] We here reproduce the original definition proposed by Granger (1981). Different definitions have been proposed, see inter alia Escribano (1987), allowing the definition to cover cases which do not fit in the ARMA framework. They more or less all reduce to the idea that an $I(d)$ process is one which requires d differences to become weakly stationary

the normalisation by $\sqrt{T}$, where T is the sample size, is no longer sufficient to guarantee the applicability of central limit theorems. For example, from (2.15), we see that the order of magnitude[8] of x_t^2 is now $O_p(T^2)$ instead of $O_p(T)$ which applies for stationary processes. Similar changes in the orders of magnitude hold for most moment and cross moment expressions with $I(1)$ processes. In particular, most of these do not converge to constants but rather to random variables. This shows the intuition behind the modification of the asymptotic theory that is required in the case of integrated processes for estimation, tests and inference (see Park and Phillips, 1988, 1989, for a detailed and general theoretical presentation and Dolado, 1991, for a good overview of the basic concepts and tools). It also explains theoretically (see Phillips, 1986) the well-known Monte Carlo study of Granger and Newbold (1974) who pointed out the danger of spurious regressions[9] when the variables in the regressions are independent integrated processes.

It may be useful to use a graphical illustration to insist upon the difference between stationary and integrated processes. Let us consider the convergence of the sample mean, i.e. $T^{-1}\sum_{i=1}^{T} x_t$, to its true constant value which is known to apply for stationary processes at the rate $T^{1/2}$ (see Gourieroux et Monfort, 1989, Fuller, 1976). Figure 2.1 reports the evolution of $\sqrt{T}T^{-1}\sum_{i=1}^{T} x_t = T^{-1/2}\sum_{i=1}^{T} x_t$ for T going from 1 to 3000 and for a first order autoregressive stationary process given by $x_t = 0.50x_{t-1} + \nu_t$ where ν_t is n.i.i.d. with zero mean and unit variance.

As expected from theory, the quantity converges relatively rapidly to its true value zero. Figure 2.2 reports the same quantity, i.e. $T^{-1/2}\sum_{i=1}^{T} x_t$, but now for an $I(1)$ process given by $x_t = x_{t-1} + \nu_t$ with the same distributional assumptions on ν_t. Note the difference in the scale of the ordinate. From the picture, it is clear that the convergence to a constant no longer holds. Note however that a similar picture emerges (Figure 2.3) if we consider a stationary process with a root near unity $x_t = 0.99x_{t-1} + \nu_t$ for which the

[8] Consider a k-dimensional random variable X_n. We say that the sequence of random variables $\{X_n\}$ is at most of order in probability a_n, denoted by $X_n = O_p(a_n)$, if for any $\epsilon > 0$ there exists a real positive number M_ϵ such that $P\{| X_{jn} | \geq M_\epsilon a_n\} \leq \epsilon$, $\forall n$, with $j = 1, 2, \ldots, k$, where $\{a_n\}$, is a sequence of positive constants (see Fuller, 1976, p.181). The interested reader is referred to Fuller (1976) or White (1984) for further details. The notation O_p generalises the order of magnitude O for real numbers to random variables.

[9] Notice however that there is no strict need to have integrated processes in order to find spurious regression. As noted by Palm and Sneek (1984), it suffices that the variables entering the regression model are independent but serially correlated.

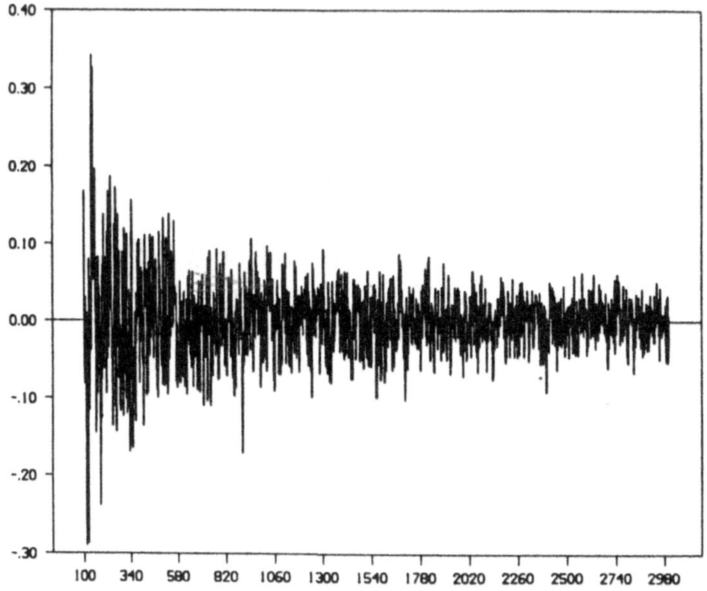

Figure 2.1: Stationary process

Figure 2.2: $I(1)$ process

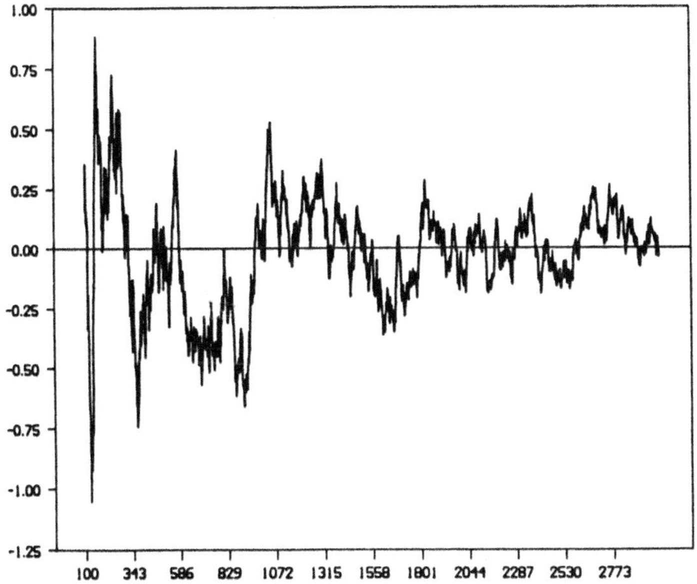

Figure 2.3: near-$I(1)$ process

behaviour of the sample moment can be more adequately represented by that of an integrated process than by that of a stationary process (see Campbell and Perron, 1992). This illustrates graphically the behaviour of sample moments of near-integrated processes as discussed in Phillips (1988b).

Similar results are observed for higher order sample moments. Overall, as the convergence of the sample moments to constants is no longer verified, statistical theory usually referred to in the case of stationary processes no longer holds. An interesting case arises, however, when linear combinations of integrated processes annihilate the individual nonstationarities. This is exactly the case when the variables are **cointegrated**, i.e. when they move stochastically together over time towards a stationary "long run" equilibrium.

Cointegration thus allows for the description of stable long run stationary relations between integrated variables.

Following Engle and Granger (1987) we say that

Definition 3
The components of the vector stochastic process x_t *are* **cointegrated of order** d, b, *denoted by* $x_t \sim CI(d, b)$ *if:*

1. *all components of* x_t *are* $I(d)$;

2. *there exists a vector* $\alpha(\neq 0)$ *such that* : $z_t = \alpha' x_t \sim I(d - b), b \neq 0$.

The vector α *is then called the* **cointegrating vector**.

The most frequent case is that of $d = b = 1$, i.e. each varaibles in x_t is $I(1)$ but a linear combination of them (z_t) is stationary. When an economicallly meaningful interpretation can be attached to this linear combination it implies that, under the hypothesis of cointegration, the components of x_t do not drift too much apart and are tied together by some long run equilibrium relation. Again a graphical illustration may prove interesting. Figure 1.4 reports the evolution of two independent $I(1)$ series with non-zero drift terms ($\mu = 0.01$ and 0.02 respectively). Although statistically independent, these two series could be spuriously considered as linked by the fact that they are both functions of time due to their non-zero drifts and trending unconditional variances. Figure 1.5 presents the graphical evolution of two $I(1)$ which are now cointegrated. Their behaviour is such that although short run deviations are allowed for, they move together over time (and never drift too much apart) towards some stationary equilibrium relationship. "Equilibrium" is here interpreted as a statistical notion which does often, but not necessarily, correspond to that of economic equilibrium.

Before discussing how one can model cointegrated time series it is worth paying some attention to the question of when and why cointegration is a useful concept. Although there are an important number of situations where economic theory implicitly leads to integration and cointegration, we shall point out two major situations where cointegration is implied by theory. The most well known situation where economic theory implies the presence of unit root nonstationarity is the efficient market hypothesis, but integration and cointegration also occur in various other situations. The first one

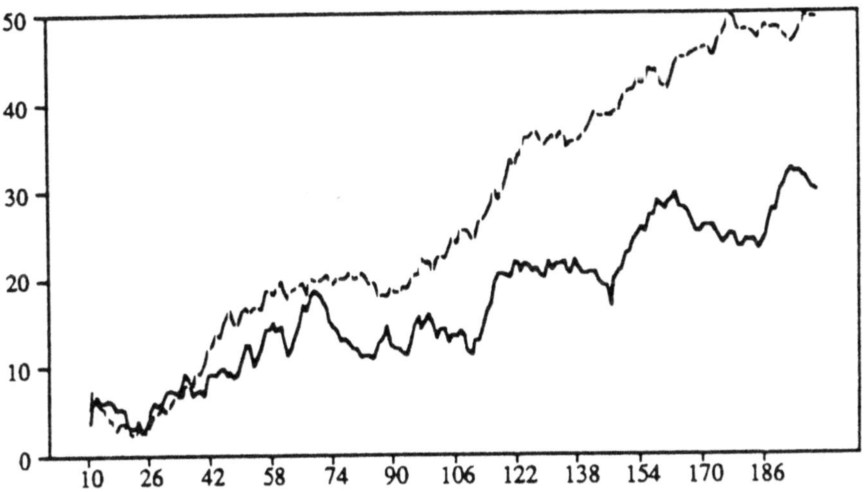

Figure 2.4: Independent $I(1)$ processes

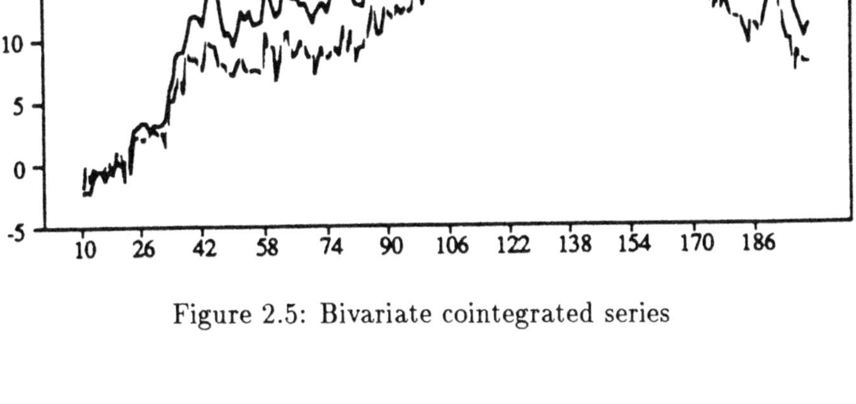

Figure 2.5: Bivariate cointegrated series

is where a long run equilibrium or steady state theory implies proportionality between two (or more) - nonstationary- economic variables by allowing these to deviate from this "equilibrium" in the short run. The most popular example is probably long run purchasing power parity which has been extensively analysed by means of cointegration techniques over the recent years (see Delhausse and Urbain, 1990, 1991, and the references cited therein). What is usually underlying is the idea that some long-run equilibrium relations are described by the theory and then the short run adjustment is similar to a servo-mechanism which is open to empirical modelling. This line of research is that of the early studies using error correction models (for example Sargan, 1964, Davidson et al. 1978). A second important field where cointegration emerges is *present value models* as pointed out in Campbell and Shiller (1987, 1988). In this case, cointegration arises from the presence of rational forecasts, instead of a result of a servo-mechanical type of causality. More precisely, if one variable is $I(1)$ and another is a rational forecast of future values of the former, then the variables will be cointegrated. The simpler example is the term structure of interest rates which implies cointegration between short and long run interest rates as far as the risk premia is stationary (see Campbell and Shiller, 1987). A further situation which has recently been analysed is that of models for forward looking optimisation agents with $I(1)$ forcing variables. This situation can then be seen as mixing the two cases just described; it arises, among others, in business cycle models (see King, Plosser and Rebelo, 1988), see also Palm and Pfann (1991). There are thus many economic arguments in favor of cointegration, besides those of pure statistical origin.

2.3 The Modelling of Cointegrated Systems

As pointed out by several authors (see Phillips, 1991b), the concept of cointegrated time series, has revolutionised both the empirical and theoretical work in time series econometrics, and this for several reasons. Firstly, cointegration provides an elegant and very easy way of modelling and estimating long run equilibrium relationships among economic variables, providing thereby a statistical framework for the investigation of economic-theoretic relationships. Secondly, it solves the problem associated with *spurious regressions* often encountered with individual integrated variables. The super consistency property demonstrated by Stock (1987) -see also Engle and Granger

(1987), Phillips and Durlauf (1986)- allows (theoretically) to estimate consistently the stationary linear combination by simple static regression without the need of modelling the short run dynamic since static ordinary least squares (OLS) estimates of cointegrating relationships are of order $O_p(T^{-1})$ instead of the usual stationary $O_p(T^{-1/2})$.

On the other hand, as pointed out by Gilbert (1986), cointegration also constitutes a formal statistical support to the empirical popular "general to specific" modelling framework which very often results in parsimonious ECMs, in an attempt to distinguish carefully between short run effects and long run effects as shown above. This link between ECM and cointegrated time series was first noticed by Granger (1981), see also Granger and Weiss (1983), and is presented and proven in Engle and Granger (1987) as the Granger Representation Theorem (GRT). Let us simply sketch some of the basic ideas underlying this theorem. Notice that we do not consider deterministic components in order to simplify the presentation. We make the following simplifying assumptions on a n-dimensional vector time series $\{x_t\}$:

Assumptions

1. *All components of x_t are $I(1)$ variables so that Δx_t is a zero mean (purely nondeterministic) weakly stationary vector stochastic process*
2. *The joint distribution of Δx_t is n dimensional multivariate Gaussian with finite mean and p.d.s. covariance matrix Σ*
3. *There exists a unique (up to a scalar factor) linear combination of the x's which is weakly stationary and denoted by $\alpha' x_t$*

From these assumptions we can write down the multivariate Wold representation of Δx_t :

$$\Delta x_t = \sum_{i=0}^{\infty} C_i \varepsilon_{t-i} = C(L)\varepsilon_t \tag{2.17}$$

with $C(L)$ a $n \times n$ matrix polynomial in the lag operator with exponentially decreasing weights (matrices) C_i. We can decompose $C(L)$ as follows

$$\sum_{i=0}^{\infty} C_i L^i = \sum_{i=0}^{\infty} C_i + (I - L) \sum_{i=1}^{\infty} (- \sum_{j=i}^{\infty} C_j) L^{i-1} \tag{2.18}$$

so that Δx_t can be rewritten as

$$\Delta x_t = C(1)\varepsilon_t + \Delta C^*(L)\varepsilon_t \tag{2.19}$$

with $C^*(L)$ is a matrix lag polynomial defined by $C^*(L) = \sum_{i=1}^{\infty} C_i^* L^{i-1}$ where $C_i^* = -\sum_{j=i}^{\infty} C_j$, $C(1) = \sum_{i=0}^{\infty} C_i$ and where it can be shown (see for example Engle and Granger, 1987) that $C^*(L)$ is invertible. If we now integrate both sides of the last equation we get

$$x_t = C(1)\sum_{s=1}^{t} \varepsilon_s + C^*(L)\varepsilon_t + (x_0 - C^*(L)\varepsilon_0) \tag{2.20}$$

which corresponds to the **Common Trends** representation of Stock and Watson (1988). If we premultiply (2.20) by α', we get

$$\alpha' x_t = \alpha' C(1)\sum_{s=1}^{t} \varepsilon_s + \alpha' C^*(L)\varepsilon_t + \alpha'(x_0 - C^*(L)\varepsilon_0)$$

From this, it is clear that for the linear combination $\alpha' x_t$ to be weakly stationary, we have to require that α lies in the left null space of $C(1)$, i.e. that $\alpha' C(1) = 0$. This gives conditions on the rank of $C(1)$ for cointegration to hold, namely that of the numbers of **common (stochastic) trends** in the system.

The **Vector AutoRegressive** form can also easily be derived from the general assumption made above by simply adding the following condition on the conditional memory of the $\{x_t\}$ process.

Assumption

4. The conditional memory of x_t is bounded to n periods.

In that case we have $E(x_t|X_{t-1}) = \sum_{i=1}^{n} A_i L^i x_t$ and $\text{Var}(x_t|X_{t-1}) = \Sigma$. Defining $\varepsilon_t = x_t - E(.|.)$, we get the **Vector AutoRegressive** representation:

$$A(L)x_t = \varepsilon_t \tag{2.21}$$

with $A_0 = I$ and $A(L) = \sum_{i=0}^{n} A_i L^i$ and ε_t is by construction normally distributed. Alternatively, and more rigorously, we could have derived the

VAR representation by inverting the matrix polynomial $C(L)$ in (2.17). Since this polynomial is singular at one value of its argument, i.e. at 1, the inverse exists at all other points. As pointed out *inter alia* in Engle and Yoo (1991), one must however be carefull in isolating the reduced rank components.[10]

From (2.17), (2.20) and (2.21) we have

$$\begin{aligned} A(L)C(L) &= \Delta I_n \\ &= (1-L)I_n \end{aligned}$$

with I_n the identity matrix, and thus

$$A(1)C(1) = 0$$

which implies that we can write $A(1) = \gamma\alpha'$ as a matrix of reduced rank. The number of cointegrating vector can thus be based on the rank of $A(1)$ or on that of $C(1)$. In particular, the analysis of the rank of A(1) gives the base of, for example, Johansen's procedure (Johansen, 1988, 1991a, Johansen and Juselius, 1990) and corresponds to the number of cointegrating vectors, while Stock and Watson (1988) base their procedure on the rank of $C(1)$, i.e. on the number of common stochastic trends in the system.

We can now come state somewhat more formally a simplified version of the Granger Representation Theorem as given for example in Engle and Yoo (1987).

Granger Representation Theorem

If the vector stochastic process x_t satisfies assumptions 1-4, then:

1. *$A(1)$ is of rank 1 and $A(0) = I_n$ in (2.21)*
2. *There exist $n \times 1$ matrices α, γ such that:*
   $\alpha' C(1) = 0$, $C(1)\gamma = 0$ *and* $A(1) = \gamma\alpha'$
3. *There exists an* **Error Correction** *representation:*

$$A^*(L)(1-L)x_t = -\gamma\alpha' x_{t-1} + \varepsilon_t$$

with $A^*(0) = I_n$.

[10] As noted by Engle and Yoo (1991), an elegant solution to this problem is to rely on Smith-McMillan forms for polynomial matrices as proposed by Yoo and used for example by Hylleberg and Mizon (1989).

Alternative proofs and various generalisations of the Granger Representation Theorem are given in Engle and Granger (1987), Hylleberg and Mizon (1989), Davidson (1991), Johansen (1991a, 1992b), Gregoire and Laroque (1991), the last two paper being concerned with the generalisation to $I(2)$ cases. A good summary can be found in Clements (1990). Note (see for example Hylleberg and Mizon, 1989) that under these assumptions, there also exists other equivalent representations for x_t, including the **Common Trends** representation (2.20) as well as the **Interim Multiplier** representation which is easily derived from the VAR model:

$$\Delta x_t = \sum_{i=1}^{n-1} \Gamma_i \Delta x_{t-i} + \Gamma_n x_{t-n} + \varepsilon_t$$

where $\Gamma_i = -I + \sum_{i=0}^{n-1} A_i$.

Notice that in this study, we shall only be concerned with the ECM and the interim multiplier representation which only differs from each other in the way the short run dynamic response parameters are interpreted.

When series are cointegrated, they thus admit (and the converse also holds) an ECM representation of the general form

$$A^*(L)(1-L)x_t = -\gamma u_{t-1} + C^+(L)\varepsilon_t \tag{2.22}$$

where $A^*(L)$ is a matrix polynomial in the lag operator L, $u_{t-1} = \alpha' x_{t-1}$ is the equilibrium errors and $C^+(L)\varepsilon_t$ is a MA polynomial which equals the unit matrix I_n if the vector time series x_t can be represented by a finite VAR model.

Since the cointegrating relationship can (at least theoretically) be consistently estimated by a static regression, Engle and Granger (1987) propose a two-step approach where the long run relations are first estimated in a static regression and in a second step, an ECM is formulated using the residual from the first step as equilibrium errors. However, it has become rapidly apparent that the Engle and Granger (1987) two-step approach, although extremely simple and appealing for empirical applications, did suffer from several drawbacks. Among these we may mention:

- the substantial finite sample biases (see Banerjee et al. 1986, Boswijk, 1989)

- the implicit common factor restriction implied by static cointegration regression models (Dolado et al. 1989, Kremers et al. 1992)

- the impossibility of identifying more than one cointegrated vector among a k dimensional set of variables with $k > 2$, for which we may naturally find at most $k - 1$ cointegrating relationships

- the need of non-parametric correction (see Phillips and Hansen, 1990) or model augmentation (see inter alia Loretan and Phillips, 1991, Stock and Watson, 1991) for conducting inference on the cointegrating relationships by means of χ^2 statistics.

As a response to these limitations and drawbacks, numerous studies have proposed alternative modelling approaches and methods for the estimation of cointegrating relationships among nonstationary economic variables. These include modified single equation procedures (Phillips and Hansen, 1990, Park, 1992), various multivariate approaches as well as the adjunction of a third step in order to tackle the efficiency problem (see Engle and Yoo, 1991). Among the various multivariate approaches, we can mention Stock and Watson (1988) who work on the *Common Trends* representation of a cointegrated system. In order to test the hypothesis that there are r cointegrating vectors against the hypothesis of $r + 1$ cointegrating vectors, they check whether the first $n - r$ eigenvalues of the first order serial correlation matrix (from the residual of a principal components analysis) are unity. Phillips and Ouliaris (1988) investigate the rank deficiency of the spectral density matrix at zero frequency, while Johansen (1988), Johansen and Juselius (1990) investigate the rank deficiency of the matrix of long run coefficients in a reparametrised VAR model. These multivariate approaches usually follow from particular representation of cointegrated systems or from the statistical properties of multivariate models with integrated processes. These properties are extensively presented in Park and Phillips (1988, 1989) who give the foundation of a general asymptotic theory for multivariate regression models with nonstationary integrated processes.

Among the implications of cointegration in terms of model representation, the ECM has become a very popular dynamic specification both in a univariate framework (following the Hendry tradition) but also in vector autoregressive form due to the contribution of Johansen (1988). Taking a finite order Gaussian VAR model as a starting point, Johansen (1988, 1990a,

1991a) and Johansen and Juselius (1990) have proposed a maximum likelihood approach for both the problem of estimating and testing the number of cointegrating relationships among the components of a k-vector x_t of variables. Since the Johansen method will be extensively used in the empirical analysis of this work we present it in more details. Consider a VAR(p) representation of the $k \times 1$ vector time series $\{x_t\}$:

$$x_t = \sum_{i=1}^{p} \Pi_i x_{t-i} + \mu + \varepsilon_t$$

and reparametrise it in vector autoregressive error correction form:

$$\Delta x_t = \sum_{i=1}^{p-1} \Gamma_i \Delta x_{t-i} + \Gamma_p x_{t-p} + \mu + \varepsilon_t \tag{2.23}$$

where ε_t denotes a k-dimensional i.i.d normal variate with mean zero and non-singular, p.d.s. covariance matrix Σ and μ is vector of constant terms. It is assumed that the roots of the implicit characteristic polynomial are outside or on the unit circle. $\Gamma_i = -I + \Pi_1 + \ldots + \Pi_i$ with $i = 1, \ldots, p$. Γ_i are here *interim* multipliers.[11]

- If Γ_p has full rank, then x_t is stationary and all the components are $I(0)$.
- If rank$(\Gamma_p) = 0$, then all components of x_t are $I(1)$ and the model is a VAR model in first differences.
- The interesting cases arise when rank$(\Gamma_p) = r < k$ in which case there are $k - r$ nonstationary linear combinations which have the effects of *common trends* and r linear combinations which are stationary, i.e. there are r *cointegrating* relationships. Γ_p can then be written as $\gamma\alpha'$ where both γ and α are $(k \times r)$ matrices of full column rank. The r first rows of α' are the r cointegrating vectors while the elements of γ are the *factor loadings*, i.e. the weights of the different cointegrating vectors in the different equations.

Note that the cointegrating matrix is not identified, but the space spanned by its columns is: for any non-singular $(r \times r)$ matrix P we can define $\alpha^* = \alpha P'$

[11] This representation is therefore referred to as the *interim representation* by Hylleberg and Mizon (1989).

and $\gamma^* = \gamma P^{-1}$ so that $\gamma^* \alpha^{*'} = \gamma\alpha'$. Provided that none of the elements of x_t are integrated of an order higher than one,[12] the maximum likelihood estimate of a basis of the cointegrating space, denoted by sp(α), is given by the empirical canonical variates of x_{t-p} with respect to Δx_t corrected for the short run dynamic and the deterministic components. The number of cointegrating relationships is given by the number of significant canonical correlations. Their significance can be tested by means of a sequence of Likelihood Ratio (LR) tests whose limiting distribution is expressed in terms of vector Brownian motions (see Johansen, 1988, 1990a, 1991a). Two possible test statistics can be used for the hypothesis of the existence of r cointegrating vectors. First the so-called *trace test*, i.e. the LR test statistic for the hypothesis that there are at most r cointegrating vectors, given by:

$$-2\log(Q) = -T \sum_{i=r+1}^{k} \log(1-\lambda_i)$$

where $\lambda_{r+1}, ..., \lambda_k$ are the $k-r$ smallest squared canonical correlations. An alternative is to use the *maximum eigenvalue test* which seeks to compare the hypothesis of r cointegrating vectors against that of $r-1$ cointegrating vectors. The LR test statistic for this hypothesis is given by:

$$-2\log(Q) = -T\log(1-\lambda_r)$$

The limiting distribution of $-2\log(Q)$, which is a function of a $k-r$ dimensional vector Brownian motion, is not independent of the unknown drift term (Johansen and Juselius, 1990). Critical values have been tabulated by Johansen and Juselius (1990) for various hypothesis concerning the behaviour of the deterministic components. Note that Johansen's (1990) approach tests for "deterministic cointegration" in the terminology of Ogaki and Park (1990).[13] This can be seen by using the following representation of x_t, which derives from the Granger Representation theorem

$$x_t = x_0 + C(1)\sum_{i=1}^{t} e_i - C(1)\mu t + C^+(L)\varepsilon_t$$

[12] For the analysis of $I(2)$ models, see Johansen (1990c), Gregoire and Laroque (1991).

[13] Deterministic cointegration can be intuitively understood as a linear combination of variables which annihilate both the linear deterministic and stochastic trend components, see Ogaki and Park (1990).

where $C(1)$ is defined from the Wold representation of Δx_t as the impact coefficient from $\Delta x_t = C(L)(e_t - \mu)$, and $C^+(L)$ is obtained by simply partitioning the lag polynomial matrix $C(L)$ as

$$C(L) = C(1) + (1 - L)C^+(L)$$

e.g. by expanding $C(L)$ around $L = 1$. We first note, from the Granger Representation Theorem, that α' lies in the left null space of $C(1)$ so that the cointegrating vectors annihilate both the stochastic ($\sum_{i=1}^{t} \varepsilon_i$) and the deterministic (μt) trends. On the other hand, the right null space of $C(1)$ is given by $sp(\gamma)$. If μ lies in $sp(\gamma)$ then the drift cancels out and the error correction term becomes

$$\gamma\alpha' x_{t-n} + \gamma(\gamma'\gamma)^{-1}\gamma'\mu$$

If μ does not belong to $sp(\gamma)$ then the non zero drift term is given by

$$[I_k - \gamma(\gamma'\gamma)^{-1}\gamma']\mu$$

An important point is therefore that Johansen's procedure does not yield a similar test since the asymptotic distribution depends on the unknown μ as is usual for most unit root tests (see inter alia Evans and Savin, 1984). The distributions of both the trace and the eigenvalue tests in the case without drift have broader tails than in the case with drift. Empirical applications should therefore carefully investigate the presence or absence of trends in the long run. A recent extension to models with quadratic trends is proposed in Johansen (1991c). Once the number of cointegrating relationships has been determined, it is possible to test particular hypotheses on γ and/or α using LR tests which have the usual limiting χ^2 distribution (see Johansen, 1988, 1991a, Johansen and Juselius, 1990, 1992a).

Until now, we have only considered what could be called *statistical representations* of multivariate cointegrated times series. In the econometric jargon, these models are seen as *reduced form* models. Then the obvious question is how could we eventually conciliate statistical models with structural econometric models which impose more "structure" in the empirical analysis.

In a stationary world, Monfort and Rabemanajara (1990) present a sequential procedure which starts from a VAR model in first differences, and then sequentially tests for Granger non-causality, predeterminedness, identification and so on, in order to arrive at a well defined structural model.

A similar approach was advocated by Palm (1983) in his SEMSTA framework. He proposed to take the VAR model as the maintained general model against which subsequent simplification will be tested, see also Palm and Pfann (1990) for an extension to nonstationary series.

In a nonstationary $I(1)$ framework, this will depend on the purpose of the analysis. In particular, the Johansen framework can easily be modified in order to obtain an identified long run structure (see Johansen and Juselius, 1992a, 1992b) by imposing a posteriori identifying restrictions on the various cointegrating vectors. In analogy with the traditional simultaneous equation model, (as discussed in Johansen and Juselius,1992b), we therefore have to impose (in addition to the normalisation of the vectors) $r-1$ restrictions of the form $R_i\alpha_i = 0$ on each cointegrating vector (α_i) in order for the system of long run (stationary) relationships to be identified, with R_i a $r-1 \times k$ matrix of known elements. In other words, if we define a matrix B_i orthogonal to R_i, i.e. $B_i = R'_{i\perp}$, we require the ith cointegrating vector to satisfy $\alpha_i = B_i\phi_i$ where ϕ_i is composed of unknown parameters. The rank condition for identification requires

$$\text{rank}(R_i\alpha) \geq r-1 \qquad i = 1, ..., r$$

In the equality holds, then just identification is achieved, otherwise the system of long run relationships will overidentified.

When the short run structure has to be jointly investigated, it might be preferable to follow the approach recently advocated by Hendry and Mizon (1990), see also Clements and Mizon (1991), for the modelling of structural econometric models[14] by starting from a VAR model which is reparametrised into a SEM. The encompassing properties of the latter are then tested against the congruent unrestricted VAR model which is, following Spanos (1990a), a characterisation of the Haavelmo distribution against which overidentifying restrictions will be tested. In fact their framework fits in the so-called encompassing analysis of Mizon (1984). The idea is first to perform a cointegration analysis using Johansen's approach, and once the cointegrating relationships have been determined and properly normalised, to divide the analysis into three major steps. First, the unrestricted VAR (UVAR) model is reparametrised in order to be an acceptable constant parameter repre-

[14]There is no unanimous acceptance of the structural character of these models, see for example the discussion in Sims (1991) who argues that these are reparametrised multiple times series (reduced form) models more than structural models.

sentation of the data. The analysis is then mapped from the space of $I(1)$ variables to the space of $I(0)$ variables where a parsimonious VAR (PVAR) model is formulated. A traditional SEM is specified and the final (complete) SEM, estimated by full information maximum likelihood (FIML), compared to the PVAR. The overidentifying restriction implied by the SEM are tested against the PVAR and checked for their validity.

A simpler two-step approach is proposed by Chow (1990) who reanalyses the accelerator-multiplier model in the light of cointegration. He first specifies a structural simultaneous equation model with clear identifying restrictions and, since the data are nonstationary, investigates the cointegrating properties using Johansen's method.

These are naturally complete models where all variables are considered as endogenous and where no exogenous variables are explicitly allowed for.

We must also mention a different route of research which has been taken by several authors, including Park (1990) and Saikkonen (1991b). These authors consider systems of (static) cointegration regressions and develop non-parametric corrections, in the spirit of Phillips' work, to ensure efficient inference. We shall not consider this line of research since our interest lies in empirical dynamic models with meaningful short run dynamic coefficients, in the spirit of Hendry and Richard's (1982, 1983) framework.

2.4 Cointegration and Conditional Sub-systems

Another way of analysing cointegration in relation to structural SEMs has been proposed by several authors who investigate cointegration in open systems, i.e. in systems where some variables are considered as strongly, or at least weakly exogenous for the parameters of interest.

The motivation for the use of conditional sub-systems stems from the traditional way one usually partitions the set of variables under investigation between so-called endogenous variables and those whose generating model will not be taken into account, the so-called "exogenous" variables. Although proceeding this way implicitly involves the risk of imposing invalid exogeneity assumptions, it nevertheless has the advantage that the analysis is conducted in open systems where some variables may be added without affecting the degrees of freedom and the size of the model. VAR models or closed-type approaches do *marginalise* w.r.t. a number of variables and their past, while open systems do *condition* on particular variables. The potential loss of information is different and crucially depends on exogeneity and non-causality properties (see for example Florens and Mouchart, 1985). Which one is the more likely to create inferential problem depends clearly on the study.

It must also been pointed out that when formulating a VAR model, the selection of the variables is usually influenced by economic theory or previous empirical regularities or findings. Typical examples are the studies of Johansen and Juselius (1990) and Hendry and Mizon (1990) where the variables selected are those usually believed to enter a long run money demand equation. Therefore, it is clear that the variables appearing in the model, if eventually sufficient to model money demand, will in turn not be sufficient to explain the evolution, of both long and short run, of the remaining variables in the system. This probably explains why some studies (Clements and Mizon, 1991, Hendry and Mizon, 1990) based on full models had to include several dummies to get parameter constancy and normality of residuals (see also the comment by Kirchgassner, 1991). These dummies might reflect the insufficiency of the conditioning information set which could be expanded to include more relevant variables eventually explaining these *structural* events. In this respect, one may expect a conditional analysis (implicitly under limited information) to be more robust against misspecification errors in the remaining equations. Moreover, as practical empirical applications clearly show (see Hendry and Mizon, 1990, Clements and Mizon, 1991), although the empirical analysis starts from a complete VAR model which is transformed into a complete SEM, the economic interest lies only in a subset of equations so that sub-system analysis could be equally appropriate.

Conditional models, on the other hand, run the risk of imposing wrong exogeneity assumptions but easily accommodate variables to appear only in some equation without requiring to be explicitly modelled. As pointed out by Kirchgassner (1991), see also Malinvaud (1981), *"often, we have to sacrifice such a more detailed description in order to get more variables in the system..."*. Basically, the problem is very similar to the choice between limited versus full information analysis of simultaneous equation models where exogeneity and overidentification may play an important role.

Conditional systems are extensively analysed in a cointegration framework by Davidson (1986), Davidson and Hall (1990, 1991) who proposed a typology of open systems which exhibit cointegration with the possibility of $I(d)$ variables with $d > 1$. The interest of Davidson (1986) lies however more specially in the mathematical structure of cointegrated systems than in deriving an operational procedure for estimation and tests. Therefore we refer the reader to Davidson (1991), Davidson and Hall (1990, 1991) and Clements (1990) for further details. Note also the interesting contribution of Dinter (1990) who investigates the asymptotic behaviour of estimators in ADL systems with $I(1)$ variables where the cointegrating relationships are estimated by ML or OLS.

Another interesting way of analyzing open cointegrated systems has been proposed by Dolado et al. (1989), Kremers et al. (1992) and Peter Boswijk (1989, 1990, 1991a, 1992a,b). Their contributions are more directed towards the elaboration of efficient estimation and testing procedures for the detection of cointegrating relationships among several economic time series in a *conditional model* (Dolado et al., 1989, Boswijk, 1989, Kremers et al. 1992) or in a *structural model* (Boswijk, 1991a, 1992a,b), i.e. in a conditional model where some identifying information is brought into the long run specification of the model. The particular advantages are threefold, at least: under the testable hypothesis of weak exogeneity (see Chapter 3, Boswijk, 1991a, 1992b, Johansen, 1992a) the approach is equivalent to ML estimation and inference on the long run parameters can be conducted by means of traditional χ^2 statistics; as pointed out above it allows economically meaningful restrictions to be easily incorporated on the long run behaviour while further overidentifying restrictions also can be introduced at a later stage.

Consider the following structural error correction model:[15]

$$\Gamma \Delta y_t = \kappa + \Pi_0 \Delta z_t + \lambda \alpha' x_{t-1} + \sum_{i=1}^{p-1} (\Phi_i \Delta y_{t-i} + \Pi_i \Delta z_{t-i}) + \eta_t \qquad (2.24)$$

where Γ is $g \times g$ nonsingular matrix with unit diagonal elements.[16] y_t is here a $g \times 1$ vector of endogenous variables, z_t a $(k-g) \times 1$ vector of weakly exogenous variables and η_t is a martingale difference sequence with respect

[15]Its derivation can be obtained by premultiplying appropriately a conditional ECM; for a more complete derivation of this specification, see Chapter 3 or Boswijk (1991a).

[16]and possibly other identifying restrictions.

to the σ-field generated by the set $(z_t, x_{t-1}, ...)$ and λ is a *diagonal* matrix of structural error correction coefficients. The rank of the cointegrating matrix, here denoted by α' is equal to r which in turn is at most equal to g so that all cointegrating relationships can be identified from the conditional model of y_t given z_t. Moreover, following Boswijk (1991a) one assumes that the rank of λ is equal to r so that no cointegrating relationships drops out of the conditional model. At this stage the cointegration relationships and the loading vectors are not yet identified. For that purpose, Boswijk (1991a) proposes[17] to identify the *separate* long run relationships by imposing restrictions of the type

$$R_i \alpha_i = e_i \qquad i = 1, \ldots, r$$

where R_i and e_i are matrices with a priori known coefficients such as ones and zeros. They are respectively of size $h_i \times k$ and $h_i \times 1$. α_i on the other hand is simply the ith column of α, i.e. the ith cointegrating vector.

Given that η_t is assumed to be uncorrelated, identification can be viewed in a similar way as the traditional identification problem of simultaneous static equation models (see inter alia Hsiao, 1984). For the model to be just identified, we require $r = g$, e.g. the number of cointegrating relationships to be equal to the number of endogenous variables. Finally, defining a $k \times (k - h_i)$ matrix Q such that $R_i Q_i = 0$ and a $k-$vector q_i such that $R_i q_i = e_i$ we can express α_i as

$$\alpha_i = Q_i \theta_i + q_i \tag{2.25}$$

so that the unrestricted coefficients of α_i are contained in the $(k - h_i) \times 1$ vector θ_i.

Under these (just)identifying restrictions, Boswijk (1991a) proposes to test for the significance of the cointegrating relationships using a simple Wald test for the significance of the error correction term, based on OLS estimation of the model (when $g = 1$) or instrumental variables (IV) method when $g > 1$. Notice that, for testing the presence of three cointegrating vectors, this approach requires the use of three Wald tests since Boswijk's approach implicitly looks at the *direction* of the rank deficiency of the long run coefficient matrix. This way of testing for the presence of cointegrating relationships is a natural generalisation of the simple $t - test$ on the ECM

[17]See also Park(1990).

term proposed by Dolado et al. (1989) and Kremers et al. (1992) for the case of bivariate relationships with known cointegration relations. The limiting distribution of the Wald test is not of χ^2 type and is expressed as a functional of a vector Brownian motion whose critical values are tabulated in Boswijk (1989, 1991a). As happens with the Johansen procedure, the critical values depend on both the potential trends and drift in the conditional representation. In particular, when the exogenous variables have linear trend components, the limiting distribution depends on unknown nuisance parameters unless a deterministic trend term is added to the regressors. Once the number of cointegrating relationships has been established, their estimates are obtained by using indirect estimators, equivalent to ML estimation under weak exogeneity. The properties of the Wald test and the ILS-estimation (for a unique cointegrating vector) seem reliable in the simulation outcome reported by Boswijk (1989) and Franses and Boswijk (1992), at least compared to the Engle and Granger (1987) two-step approach and Johansen's ML approach.

2.5 Error Correction Models

One common feature of many of the above mentioned approaches is that they usually rely on some form of error correction model in their empirical implementation.[18] Different interpretations of the ECM have been given in the literature and it might be useful to clarify this by a brief overview of the different interpretations.

Before the emergence of the cointegration literature, ECMs were already very popular in applied econometrics, due to the work of Sargan (1964) and especially the consumption study of Davidson, Hendry, Yeo and Sbra (1978). Recently, Alogoskoufis and Smith (1991) have provided a tentative review of ECMs to which we refer for further details on these precursors. Consider a

[18] As correctly argued by Gourieroux and Monfort (1990), one should more exactly use the label *error correction form* since almost all dynamic models may be rewritten in error correction format without imposing particular constraints on the parameters. Notice that in the early application of single equation error correction models, long run proportionality was often assumed (see Davidson et al., 1978) for theoretical convenience. As pointed out in Banerjee et al. (1990) this is just a restricted ECM, sometimes labelled the homogeneous ECM.

simple single equation ECM:

$$\Delta y_t = \beta_1 \Delta x_t + \beta_2(y_{t-1} - x_{t-1}) + \varepsilon_t \tag{2.26}$$

where we have, for notational convenience, assumed a unit long run coefficient.

The major problem that one encounters is that a model similar to this ECM can have different interpretations which are very difficult to distinguish unless the models are not completed by auxiliary assumptions or by a generating model for the exogenous variables. In order to minimise the overlap with Alogoskoufis and Smith (1991) - hereafter AS-, we shall base our brief presentation on the alternative interpretation of (single equation) conditional dynamic models discussed in Hendry, Pagan and Sargan (1984).

- The first interpretation is in terms of a *regression model* where the parameters are defined according to

$$E(\Delta y_t | \Delta x_t, y_{t-1}, x_{t-1}) = \beta_1 \Delta x_t + \beta_2(y_{t-1} - x_{t-1}). \tag{2.27}$$

  where $\varepsilon_t = \Delta y_t - E(.|.)$ so that by construction the r.h.s. variables are orthogonal to ε_t. The question arises of the interpretation of the parameters of such models and more specifically their potential relation (or lack thereof) with the *theoretical parameters of interest*.[19] This is a main criticism of AS who argue that if we do not add assumptions concerning for example expectation behaviour, then the parameters of such an ECM will likely be a mixture of equilibrium, expectation and adjustment parameters, in which case the Lucas critique is certainly applicable.

- The latter potential problem naturally does not arise when the single equation ECM is derived from a well defined dynamic theory in which case the parameters directly characterise how economic agents form plans and expectations. Such a derivation is proposed, for example by Nickell (1985) and discussed in AS. The ECM is then viewed as an optimal adjustment rule deriving from tâtonnement-type processes or from optimal responses of agents in an intertemporal context. Examples following this approach include Nickell (1985), Domowitz and

[19] Usually these are related to economically meaningful notions.

Hakkio (1990), Keil and Richardson (1990). Most of these studies base their derivation on quadratic loss functions which penalise the deviation from equilibrium and adjustment. Assuming, for example, a second order autoregressive process with a single unit root for the target, Nickell (1985) shows that the ECM arises as a solution to the minimisation of a forward looking quadratic costs function. Note however that a potential source of problems comes from the models for the forcing variables which are often based on simple univariate time series models, see AS, p.119. Misspecification of these by, for example, not taking into account the cointegration properties, might eventually affect the empirical outcome.

- A final interpretation which is proposed inter alia by Lubrano, Pierce and Richard (1986) is to interpret the single equation in terms of behavioral relations relating the expectation of $y_t|X_{t-1}$ and $z_t|X_{t-1}$. This places the analysis in the realm of the Hendry and Richard (1982, 1983) framework and has been strongly advocated by Hendry (1988) as an alternative to expectation based behaviour. As misspecification tests are an important feature of this modelling framework, it can provide an interesting alternative to theoretical-based models. In fact, the Lucas critique, which implicitly underlies some motivation of these, can be tested on the final parametrisation along with other useful design criteria. In particular Hendry (1988) shows that if a conditional feedback model -such as the single equation ECM- has constant parameters while the marginal model for x_t has nonconstant parameters, then the forward expectation hypothesis cannot hold.

Note that ECMs can be derived both in a stationary framework and in a nonstationary context. In the latter case, the cointegrating properties of the data have to be accounted for, and potentially lead to higher dimensional models unless exogeneity assumptions are verified.

2.6 Conclusions

Numerous estimators and procedures has been proposed over the last few years for the estimation and detection of cointegrating vectors among a set of nonstationary integrated time series. While they differ from the representation of the cointegrated system on which they are based, there are some other important differences to stress.

Generally, system approaches are superior to those based on sub-systems or single equations, as pointed out by Phillips (1991a) and Phillips and Loretan (1991). These system approaches do not require any exogeneity assumptions, and when they do not *estimate* the unit roots in the system, but *impose* them as in the Johansen (1988, 1991a) framework, then usual χ^2 statistics are applicable since the limiting distribution of the cointegrating vector estimator is mixed normal (see for example Phillips, 1991a). This is however also true for appropriate conditional models or augmented or corrected single equation models.

On the other hand, full-systems have their drawbacks such as dimensional problems. Moreover, if both the long and short run have to be modelled in an economically meaningful way, it usually leads to two-step approaches whose small sample properties have not been extensively worked out yet. From an empirical point of view, conditional representations of cointegrated systems are thus serious candidates. Their validity however relies on the correctness of the exogeneity assumptions so that it appears essential to state clearly under what conditions a given variable can be treated as exogenous in this framework.

Chapter 3

Weak Exogeneity in Error Correction Models

As has been pointed out in the previous chapter, the choice between a full system approach and a sub-system approach is not clear cut in the case of cointegrated systems. Both have advantages and defects. Conditional sub-systems, in particular, necessitate exogeneity assumptions which, if not fulfilled, may induce loss of efficiency and consistency -as in the usual stationary framework- but in the case of cointegrated systems also imply a loss of the mixed normal limiting distribution of the cointegrating vector estimators necessitating therefore semi-parametric corrections, model augmentation or two-step approaches (see Phillips and Hansen, 1990, Phillips, 1991, Phillips and Loretan, 1991, Saikkonen, 1991, Stock and Watson, 1991, Boswijk, 1992a,b).

The concept of exogeneity is thus here too of crucial importance for conducting valid inference in (conditional) cointegrated sub-systems. In the recent applied econometric literature, the error correction model (ECM) has become a very popular dynamic specification; both in a multivariate framework and in a single equation framework as pointed out in Chapter 2. In the latter case, exogeneity assumptions are implicitly made on the right hand side variables and the analysis is conducted in limited information.

The purpose of this chapter[1] is to point out that the use of a single equation ECM implicitly requires important hypotheses both upon the long run properties and on the weak exogeneity status of the various variables. This may have particular important effects when weak exogeneity is tested using traditional orthogonality tests within a limited information framework.

The organisation of this chapter is as follows. The first section provides the definition of weak exogeneity and the empirical motivation of this study. Section 2 considers the derivation of the ECM formulation as proposed by Engle and Granger (1987) in the modelling framework advocated by Hendry and Richard (1982, 1983). We derive the single equation ECM in reduced form and point out the conditions under which it is appropriate to assume weak exogeneity, and hence to perform inference within a single equation (conditional) ECM formulation. Section 3 reports similar results for the ECM in structural form. In Section 4 we discuss briefly how to check for weak exogeneity in an ECM. Section 5 investigates several empirical studies. Some conclusions are drawn in the last section.

3.1 Weak Exogeneity

In any empirical model building of economic processes one must first choose the information set, decide which variables will be treated as endogenous, in which equation, and which ones as "exogenous". The crucial importance of the concept of exogeneity for conducting valid inference in a conditional econometric model has been pointed out particularly well in Engle et al. (1983) who give an overview of the various previous definitions in the literature -see also the survey by Geweke (1984). Intuitively, weak exogeneity of r.h.s. variables means that explanatory variables may be treated as "fixed" for inference on parameters of particular interest, i.e. no useful sample information is lost when we condition on these variables without specifying their generating process.

[1] Most of the results reported in this chapter are taken from Urbain (1992b).

3.1.1 Definition and example

In the sequel we shall confine ourselves to the concept of weak exogeneity proposed by Richard (1980) and Engle et al. (1983), based on the work of Koopmans (1950) and on the statistical concept of classical cut as described in Barndorff-Nielsen (1978). Note that we shall always define (weak) exogeneity with respect to a set of parameters of interest denoted by ψ. Some authors, for example Holly (1985), define exogeneity of a variable in a given equation by considering that all the parameters of this equation are of interest. As this is not systematically the case we prefer to keep the notion relative to some particular parameters.

Consider a vector of observable time series x_t which we can partition as (y_t, z_t). Write the σ-field generated by $(X_0, x_1, ..., x_{t-1})$ as X_{t-1} where X_0 is a matrix of initial values. We assume that the initial conditions are given so that the entire analysis in conducted conditionally on X_0. Define the family of the joint density x_t as the P_θ density family with $\theta \in \Theta$ where Θ is the parameter space.

Assume that the process generating the T observations is represented by the joint data density:

$$D(x_1, \ldots, x_T | X_0; \theta)$$

Without loss of generality, the joint density of the complete sample (T observations) can be sequentially factorised as

$$D(x_1, \ldots, x_T | X_0; \theta) = \prod_{t=1}^{T} D(x_t | X_{t-1}; \theta)$$

Let focus attention on the conditional density function

$$D(x_t | X_{t-1}; \theta)$$

and suppose that there exists a one-to-one transformation f such that

$$f : \Theta \rightarrow \Lambda; \quad \theta \rightsquigarrow \lambda = f(\theta)$$

Partition λ as (λ_1, λ_2) and Λ as $(\Lambda_1 \times \Lambda_2)$. We shall say that λ_1 and λ_2 are variation free (or variation independent in the terminology of Barndorff-Nielsen, 1978, p.26) if $(\lambda_1, \lambda_2) \in (\Lambda_1 \times \Lambda_2)$, the product space of their respective parameter spaces; i.e. if and only if λ_1 and λ_2 are not subject to cross restrictions so that for any specific admissible value in Λ_1 for λ_1, λ_2 can take any value in Λ_2. Let λ_1 be a one-to-one function of our *parameters of interest* ψ and λ_2 be the so-called *nuisance parameters.*

Definition 1 : Weak Exogeneity (Engle et al., 1983)

Let focus attention on the conditional density function $D(x_t|X_{t-1};\theta)$ and let $x_t = (y_t, z_t)$. z_t will be said to be weakly exogenous over the sample period, for the parameter of interest ψ, if and only if there exists a reparametrisation λ of θ, with $\lambda = (\lambda_1, \lambda_2)$, such that

1. *ψ depends on λ_1 only,*
2. *$D(x_t|X_{t-1};\lambda) = D(y_t|z_t, X_{t-1};\lambda_1)D(z_t|X_{t-1};\lambda_2)$;*
3. *(λ_1, λ_2) are variation free, i.e. $(\lambda_1, \lambda_2) \in (\Lambda_1 \times \Lambda_2)$*

Note that conditions 2 and 3 correspond to the notion of "classical sequential cut" (see Engle et al., 1983). Weak exogeneity thus implies that the parameters of interest can be recovered from the conditional model only. It also implies absence of cross-restrictions between λ_1 and λ_2, and that the joint density of the entire sample can factorise as above. Note that, if we consider dynamic models, $D(.)$ are usually not densities, see Gourieroux and Monfort (1989, p.34), unless further Granger non-causality conditions are introduced.[2] Effectively, weak exogeneity does not exclude relationships between *lagged* y_t's and z_t. Notice that weak exoxegeneity plus Granger non-causality results in the notion of *strong exogeneity* which permits valid multi-step ahead prediction of y from the conditional model. In that case, only lagged values of z_t enter $D(z_t|X_{t-1};\lambda_2)$.

Example

In order to illustrate the concept of weak exogeneity and its importance for efficient inference in conditional models, we take an example where for

[2]See Gourieroux and Monfort (1989), Florens and Mouchart (1985), Holly (1985)

simplicity we assume a static Gaussian framework.[3] Consider the following bivariate model

$$y_t = \beta z_t + u_t \tag{3.1}$$

$$z_t = \alpha w_t + v_t \tag{3.2}$$

where z_t, y_t and w_t are scalar time series, w_t is strongly exogenous and z_t is the variable whose exogeneity is questioned. Equation (3.1) is assumed to represent a structural relation between y_t and z_t. The parameter of theoretical interest (i.e. ψ) is given by β. We also assume that the bivariate process $(u_t, v_t)'$ is identically and independently normally distributed with a zero mean vector and p.d.s. covariance matrix Σ given by

$$\Sigma = \begin{pmatrix} \sigma_{11} & \sigma_{12} \\ \sigma_{21} & \sigma_{22} \end{pmatrix}$$

The joint distribution of $(y_t \; z_t)'$ is :

$$\begin{matrix} y_t \\ z_t \end{matrix} \mid w_t \sim N\left[\begin{pmatrix} \beta \alpha w_t \\ \alpha w_t \end{pmatrix}, \begin{pmatrix} \sigma_{11} + 2\beta\sigma_{12} + \beta^2\sigma_{22} & \sigma_{12} + \beta\sigma_{22} \\ \sigma_{12} + \beta\sigma_{22} & \sigma_{22} \end{pmatrix} \right]$$

We can factorise the joint density into the product of the conditional density of $y_t|z_t$ and the marginal of z_t

$$D(y_t, z_t|w_t; \lambda) = D(y_t|z_t, w_t, \lambda_1)D(z_t|w_t, \lambda_2) \tag{3.3}$$

where

$$y_t|\, z_t, w_t \sim N[(\sigma_{12}\sigma_{22}^{-1} + \beta)z_t - \sigma_{12}\sigma_{22}^{-1}\alpha w_t, \sigma_{11} - \sigma_{12}\sigma_{22}^{-1}\sigma_{21}]$$

$$z_t|\, w_t \sim N[\alpha w_t, \sigma_{22}]$$

From the conditional density $(y_t|\, z_t, w_t)$ it is already clear that we cannot estimate β from the conditional model alone because $\sigma_{12}\sigma_{22}^{-1}$ is found in all three parameters of the conditional density. Let $\sigma_{11.2} = \sigma_{11} - \sigma_{12}\sigma_{22}^{-1}\sigma_{21}$ and $\tau = \sigma_{12}\sigma_{22}^{-1}$. The factorisation (3.3) of the joint density induces a one-to-one reparametrisation of (β, α, Σ) into $(\beta + \tau, \tau\alpha, \sigma_{11.2}, \alpha, \sigma_{22})$. If we set

[3] For similar examples see Lubrano et al. (1986), Urbain (1988). Since the model is static, the notion of weak exogeneity will naturally coincide with that of strong exogeneity.

$b = \beta + \tau$, $c = -\alpha\tau$, the conditional density of y_t given z_t can be written as the regression

$$y_t = bz_t + cw_t + \varepsilon_t \qquad \varepsilon_t \sim N(0, \sigma_{11.2}) \tag{3.4}$$

By construction (and due to the positive definiteness of Σ) the parametrisation $\lambda_1 = (b, c, \sigma_{11.2})$ and $\lambda_2 = (\alpha, \sigma_{22})$ operates a cut on the joint density. However, this does not suffice for weak exogeneity of z_t for β to hold. In particular, we see that τ appears in b, c and $\sigma_{11.2}$ so that β cannot be recovered from the conditional model only as far as $\sigma_{12} \neq 0$. In this case, the sampling distribution of the estimates of β, $\hat{\beta}$, will depend on the sampling distribution of both z_t and w_t.[4] On the other hand, if $\sigma_{12} = 0$ and if β and (α, σ_{22}) are not subject to cross-restrictions, then the conditional distribution fully characterises the structural equation and the parameter of interest can be recovered from the conditional model only. Note that the condition $\sigma_{12} = 0$ is not always necessary for weak exogeneity of z_t with respect to β. For example, if α is a known constant, it is easily seen that β can be recovered from the conditional model only for it equals $b - (c/\alpha)$. This shows that conditional independence (here $\sigma_{12} = 0$) is not always necessary for weak exogeneity to hold.

If the parameter of interest is the coefficient of z_t in the conditional model (3.4), i.e. if ψ is given by b, then z_t is weakly exogenous for b by construction. This shows that weak exogeneity is relative to a parameter of interest and not an intrinsic characteristic of a variable.

It is straightforward to extend this simple example to a dynamic model by replacing w_t by y_{t-1}. In this case the conditions of weak exogeneity differ from that of strong exogeneity since a condition of Granger non-causality has to be added in the latter case.

We can get even more insight into the *interpretation* of conditional independence (here given by $\sigma_{12} = 0$), which is a sufficient condition for weak exogeneity to hold in our example, if we rewrite (3.1) in terms of expectation (see also Lubrano et al., 1986), where the subscripts m and c respectively

[4] Assume for example that w_t is n.i.i.d.$(0, \sigma_{33})$. The conditional distribution $y_t | z_t, w_t$ will then be given by $y_t | z_t, w_t, \lambda_1 \sim N[(\beta + \tau^*)z_t, \sigma_{11} - \tau^2(\sigma_{22} + \alpha^2\sigma_{33})]$ with $\tau^* = \sigma_{12}(\sigma_{22} + \alpha^2\sigma_{33})^{-1}$. Changes in the nuisance parameters $(\alpha, \sigma_{22}, \sigma_{33})$ will then induce changes in the sampling distribution of $\hat{\beta}$.

stand for marginal and conditional.

$$\begin{aligned} E_m[y_t|w_t,.] &= \beta E_m[z_t|w_t,.] \\ &= \alpha\beta w_t \end{aligned} \tag{3.5}$$

From (3.4) we also have

$$E_c[y_t|z_t, w_t,.] = (\beta+\tau)z_t - \tau\alpha w_t \tag{3.6}$$

Since $\tau = \sigma_{12}\sigma_{22}^{-1}$, the condition $\sigma_{12} = 0$ is necessary and sufficient in this case for (3.5) to be equivalent to (3.6), although in the former the expectations are *marginal* with respect to z_t while in the latter they are *conditional* on z_t . We can reformulate (3.6) as

$$E_c[y_t|z_t, w_t,.] = \alpha\beta w_t + (\beta+\tau)(z_t - \alpha w_t)$$

where αw_t and $(z_t - \alpha w_t)$ are respectively interpreted as "anticipated" and "unanticipated" components of z_t (see Lubarno et al., 1986). Therefore, under $\sigma_{12} = 0$, the behaviour will not be modified by a possible (unexpected) change in z_t. Richard (1980) and Lubrano et al.(1986) describe this situation as one of *effective* control. If weak exogeneity holds, then efficient estimation and inference (i.e. testing) may safely be conducted by ignoring the information of the marginal model and thus by analyzing the conditional model only.

It is worth contrasting this notion of *weak exogeneity* with the more traditional concepts of *predeterminedness* and *strict exogeneity.* If we assume that we are implicitly in a complete dynamic simultaneous equation model, a variable will be strictly exogenous if it is uncorrelated with all current, past and future disturbances in the model. If it is uncorrelated with all current and future disturbances, then it will be predetermined. The distinction is thus clear and, as it appears, strict exogeneity and predeterminedness imply nothing regarding the parameters of interest. These concepts are not sufficient nor necessary for valid inference in a conditional model since a variable can be strictly exogenous or predetermined in a particular equation, but not weakly exogenous for the parameters of interest as it will appear later in this chapter -see also Engle et al. (1983).

For further discussion on these concepts we refer the reader to Engle et al. (1983) or to the interesting presentation in Ericsson (1992). The implication of these definitions with regard to the construction of parametric tests for the weak exogeneity hypothesis is discussed briefly at the beginning of the next chapter.

3.1.2 Empirical motivations

Several empirical studies which used error correction models (inter alia Lubrano, Pierce and Richard, 1986, Steel, 1987a, 1987b, Steel and Richard, 1991, Urbain, 1988) have investigated the exogeneity status of some "explanatory" variables. Steel (1987a) studied the exogeneity of disposable income in an annual nondurable consumption model for Belgium; Lubrano et al.(1986), Steel (1987b) and Steel and Richard (1991), analysed the exogeneity of interest rates and prices within a money demand model for the UK while in Urbain (1988), we analysed the exogeneity status of import prices within aggregate import demand function for several OECD countries. These studies were conducted in a limited information framework and mostly reached similar conclusions, namely the non rejection of weak (and often strong exogeneity.) This particular striking result calls for the following question:

Do we reject weak exogeneity too seldom in error correction models ?

In this chapter we focus on two possible explanations of this inferential outcome:[5]

- These studies do not consider the restrictions implied by the use of conditional error correction models.

- All these studies derived their ECMs within an incomplete simultaneous equation framework which does not explicitly take into account the nonstationarity observed in the data. When these are accounted for, the co-integration restriction implies non-linear cross equation restrictions and, consequently, weak exogeneity is destroyed in most cases.

[5]A third possible explanation is related to the sensitivity of these tests to possible misspecification of the auxiliary model (see the Monte Carlo study of Kiviet (1985) which partially address this problem, as well as the next chapter).

3.2 Reduced Form Error Correction Models

3.2.1 The error correction system in reduced form

In the line of Hendry and Richard (1982, 1983) we consider that the data, T observations on a set of k variables, denoted by $\{x_t\}$ are generated according to an unknown data generating process (DGP) represented by the joint probability density function $D(x_t|X_{t-1};\theta)$. We assume a linear normal representation of the conditional mean of the observables:

$$x_t|X_{t-1} \sim N(\mu_t, \Sigma_t) \tag{3.7}$$

where

- $X_{t-1} = (X_0, x_1, \ldots, x_{t-1})'$,
- $\mu_t = E(x_t|X_{t-1}, \theta) = \sum_{i=1}^{n} \Pi_i x_{t-i}$,
- $\Sigma_t = E[(x_t - \mu_t)((x_t - \mu_t)'|X_{t-1}, \theta]$,
- n is the lag length truncation,
- the parameters $\{\Pi_i\}$ and Σ are function of θ.

We also assume time homogeneity of the conditional process so that the Π_i's and Σ_t are time invariant (see Spanos, 1988). Defining $\varepsilon_t = x_t - \mu_t$, where ε_t is a mean innovation process with respect to X_{t-1}, gives the simple dynamic representation:

$$x_t = \sum_{i=1}^{n} \Pi_i x_{t-i} + \varepsilon_t \tag{3.8}$$

By making use of matrix polynomial notation in the lag operator L, we can rewrite (3.8) as

$$\Pi^*(L)x_t = \varepsilon_t \tag{3.9}$$

where $\Pi^*(0) = I_k$, $\Pi^*(L) = I_k - \Pi(L)$ and $\Pi(L)$ is the $(k \times k)$ matrix polynomial (of order n) in the lag operator L defined in (3.8): $\Pi(L) = \prod_{i=1}^{n} \Pi_i L^i$.

We may now state some assumptions about x_t and about the $\Pi^*(L)$ polynomial in order to consider nonstationary (co)integrated processes. We make the following two simplifying assumptions:

Assumption 1
All k components of x_t are nonstationary variables, integrated of order one, denoted by $x_t \sim I(1)$

Assumption 2
$\Pi^*(1)$ is of reduced rank $r < k$ and can hence be written as $\gamma\alpha'$ with γ and α $(k \times r)$ matrices.

These two assumptions imply (see inter alia Johansen, 1991a) that there exist stationary linear combinations of the x_t's given by $\alpha' x_t$. The rows of α are called the cointegrating vectors while the elements in γ are the factor loadings, i.e. the weights of the cointegrating vectors in the different equations. In practice we thus assume that all the roots of the determinant equation $\det(\Pi^*(L)) = 0$ are outside or at most on the unit circle. For notational convenience we omit the deterministic nonstationarities from the presentation to focus on the relations between the variables in x_t.

Throughout this chapter, we use the following decomposition of a polynomial to derive ECM forms. For a given polynomial $A(L)$ it consists of rewriting the polynomial in the following form:

$$\begin{aligned} A(L) &= \sum_{i=0}^{n} A_i L^i \\ &= \sum_{i=0}^{n} A_i L + (I - L)A_0 + (I - L)\sum_{i=2}^{n}(\sum_{j=i}^{n} -A_j)L^{i-1} \end{aligned} \quad (3.10)$$

which gives

$$A(L) = A(1)L + (I - L)A^+(L)$$

where $A_i^+ = -\sum_{j=i+1}^{n} A_j$ for $i = 0, ..., n-1$.

Recalling the VAR model (3.9), we can apply this decomposition to the matrix polynomial $\Pi^*(L)$

$$\Pi^*(L) = \Pi^*(1)L + (1 - L)\Pi^+(L)$$

with $\Pi^+(L) = \Pi_0^+ + \Pi_1^+ L + ... + \Pi_{n-1}^+ L^{n-1}$, where $\Pi_i^+ = -\sum_{j=i+1}^{n} \Pi_j^*$ for $i = 0, 1, ..., n-1$. Then (3.9) becomes

$$\Pi^+(L)(1-L)x_t = -\Pi^*(1)x_{t-1} + \varepsilon_t \tag{3.11}$$

Further noting that, from Assumption 2, $\Pi^*(1) = \gamma\alpha'$, we can rewrite (3.8) as the *reduced form Error Correction Representation* (Engle and Granger, 1987, Hylleberg and Mizon, 1989) with $\Pi^+(0) = I_k$:

$$\Pi^+(L)(1-L)x_t = -\gamma\alpha' x_{t-1} + \varepsilon_t \tag{3.12}$$

which constitutes a VAR model with non-linear restrictions among the various equations.

3.2.2 Single equation error correction model in reduced form and weak exogeneity

Assume now that we may partition $x_t' = (y_t, z_t')$ where y_t is a scalar time series and z_t is assumed to be a $(k-1) \times 1$ vector of time series. In order to derive the single equation conditional error correction model (in reduced form) we first replace the Assumption 2 by

Assumption 3
$\Pi^*(1)$ *is of rank 1 and can be written as* $\gamma\alpha'$ *where* γ *and* α *are* $k \times 1$ *vectors. We assume that we can normalise* α' *as* $(1, -\delta')$ *and partition* $\gamma' = (\gamma_1, \gamma_2')$ *where* γ_1 *is a scalar and* γ_2 *a* $(k-1) \times 1$ *vector so that*

$$\begin{aligned} \Pi^*(1) &= \gamma\alpha' \\ &= \begin{pmatrix} \gamma_1 & -\gamma_1\delta' \\ \gamma_2 & -\gamma_2\delta' \end{pmatrix} \end{aligned}$$

Notice that this simplifying assumption is stated to avoid identification problems. In other words, we assume that there exists only one cointegrating vector (up to a scalar factor) α which produces a stationary linear combination $\alpha' x_t$. The normalisation is essential for our purpose as it will allow us to identify the cointegrating vector from the conditional submodel. For a vector time series x_t with k components, there may be at most $k-1$ cointegrating vectors. So if a single equation ECM is postulated with a single error correction term, it will usually estimate a linear combination of the various cointegrating vectors (see Johansen, 1992a). This however also implies that we have to rule out the possibility that z_t alone (i.e. without y_t) is cointegrated. Note that the assumption that we may normalise α to 1 is important since if $\dot{\alpha} = (0, \delta')$, then the remaining variables in z_t are cointegrated. The generalisation to a g-dimensional vector of endogenous variables is straightforward by simply replacing Assumption 3 by the condition that there are at most g separate cointegrating vectors in which case the identification problem can be dealt with along the lines presented in Boswijk (1991a).[6] The validity of the inference conducted in single equation error correction models will then depend upon whether the conditioning variables can be considered as weakly exogenous for the parameters of interest. These will be defined later as our interest will be, among others, to point out the different conditions for weak exogeneity which emerge when the parameters of interest are changed.

Let us first derive the conditional error correction model in this framework. We may write (3.8) according to the partitioning of x_t and assume that the variables are generated according to:[7]

$$\begin{bmatrix} y_t \\ z_t \end{bmatrix} = \begin{bmatrix} \Pi_{11}(L) & \Pi_{12}(L) \\ \Pi_{21}(L) & \Pi_{22}(L) \end{bmatrix} \begin{bmatrix} y_{t-1} \\ z_{t-1} \end{bmatrix} + \begin{bmatrix} \varepsilon_{1t} \\ \varepsilon_{2t} \end{bmatrix} \tag{3.13}$$

with

$$\begin{bmatrix} \varepsilon_{1t} \\ \varepsilon_{2t} \end{bmatrix} \sim N\left[\begin{pmatrix} 0 \\ 0 \end{pmatrix}, \begin{pmatrix} \sigma_{11} & \sigma_{12} \\ \sigma_{21} & \Sigma_{22} \end{pmatrix}\right]$$

---

[6] We should also mention that the concept of *cointegrating* exogeneity proposed by Hunter (1990) is ruled out by our assumptions.

[7] Since we want to derive the single equation ECM from a VAR model, we start from this unrestricted parametrisation. We could naturally have started the analysis by supposing that our generating model is of ECM type such as (3.12). The results remain unchanged since (3.12) and (3.8) are observationally equivalent under our assumptions (see The Granger Representation Theorem in Engle and Granger, 1987). Note also that such unrestricted VAR models arise naturally if we consider the implicit steps of global conditioning and global marginalisation within the Hendry approach (see Hendry, 1987, formula (3.5) p.33).

where the lag polynomials in (3.13) are given by $\Pi_{11}(L) = \sum_{i=1}^{n-1} \Pi_{11,i}L^i$, $\Pi_{12}(L) = \sum_{i=1}^{n-1} \Pi_{12,i}L^i$ $\Pi_{21}(L) = \sum_{i=1}^{n-1} \Pi_{21,i}L^i$, $\Pi_{22}(L) = \sum_{i=1}^{n-1} \Pi_{22,i}L^i$

Assuming joint normality, the conditional density of $y_t|z_t, X_{t-1}$ is normal with conditional expectation given by

$$\begin{aligned} E(y_t|z_t, X_{t-1}, \theta_1) = \sigma_{12}\Sigma_{22}^{-1}z_t + [\Pi_{11}(L) - \sigma_{12}\Sigma_{22}^{-1}\Pi_{21}(L)]y_{t-1} \\ +[\Pi_{12}(L) - \sigma_{12}\Sigma_{22}^{-1}\Pi_{22}(L)]z_{t-1} \end{aligned}$$

and conditional variance

$$\mathrm{Var}(y_t|z_t, X_{t-1}, \theta_1) = \sigma_{11.2} = \sigma_{11} - \sigma_{12}\Sigma_{22}^{-1}\sigma_{21}$$

We define the lag polynomials

$$\begin{aligned} C(L) &= I - \sum_{i=1}^{n}(\Pi_{11,i} - \sigma_{12}\Sigma_{22}^{-1}\Pi_{21,i})L^i \\ &= I - \sum_{i=1}^{n} C_i L^i \\ &= I - C^*(L) \end{aligned} \tag{3.14}$$

and

$$\begin{aligned} Q(L) &= \sum_{i=0}^{n}(\Pi_{12,i} - \sigma_{12}\Sigma_{22}^{-1}\Pi_{22,i})L^i \\ &= \sum_{i=0}^{n} Q_i L^i \end{aligned} \tag{3.15}$$

With the notation

$$y_t - E(y_t|z_t, X_{t-1}, \theta) = \zeta_t \qquad \text{where} \quad \zeta_t \sim N(0, \sigma_{11.2})$$

the conditional density of $y_t|\, z_t, X_{t-1}$ is given by

$$y_t|\, z_t, X_{t-1} \sim N[C^*(L)y_{t1} + Q(L)z_t, \sigma_{11.2}] \tag{3.16}$$

where $C^*(L)$ and $Q(L)$ are defined in (3.14) and (3.15). The conditional model can then be written as a single equation **A**utoregressive **D**istributed **L**ag model (ADL):

$$C(L)y_t = Q(L)z_t + \zeta_t \qquad (3.17)$$

From the assumption of cointegration, the roots of the determinant equation $|(C(L)| = 0$ lie all outside the unit circle which will enable us to estimate the cointegrating vector from the conditional model only. This is the condition of stability which means that if z_t and ζ_t have been fixed at z and 0 respectively for an infinite number of periods, then the conditional model has a unique solution $y = -C(1)^{-1}Q(1)z$. Note that it is implicitly assumed in most of the single equation error correction models.

We may now use the decomposition (3.10) to write

$$C(L) = C(1)L + (1-L)C^+(L) \quad \text{and} \quad Q(L) = Q(1)L + (1-L)Q^+(L)$$

By rearranging the terms of (3.17), and writing $\Delta = 1 - L$, we obtain a **single equation conditional error correction model** with $C^+(0) = I$:

$$C^+(L)\Delta y_t = -C(1)y_{t1} + Q(1)z_{t-1} + Q^+(L)\Delta z_t + \zeta_t \qquad (3.18)$$

From Assumption 3, $\Pi^*(1) = \gamma\alpha'$ with $\alpha' = (1, -\delta')$ and $\gamma' = (\gamma_1, \gamma_2')$. Using (3.14) and (3.15) we obtain

$$C(1) = \gamma_1 - \sigma_{12}\Sigma_{22}^{-1}\gamma_2 \quad \text{and} \quad Q(1) = -\gamma_1\delta' + \sigma_{12}\Sigma_{22}^{-1}\gamma_2\delta' \qquad (3.19)$$

On the other hand the marginal density $D(z_t|X_{t-1}, \theta_2)$ is given by

$$z_t|X_{t-1} \sim N[\Pi_{21}(L)y_{t-1} + \Pi_{22}(L)z_{t-1}, \Sigma_{22}]$$

with $\Pi_{21}(1) = -\gamma_2$ and $\Pi_{22}(1) = I + \gamma_2\delta'$. In order make these vectors of parameters γ_2 and δ' explicit in the mean of the marginal density, we decompose $\Pi_{21}(L) = \Pi_{21}(1)L + (1-L)\Pi^{+}_{21}(L)$ and $\Pi_{22}(L) = \Pi_{22}(1)L + (1-L)\Pi^{+}_{22}(L)$ so that the marginal density can be rewritten as

$$\begin{aligned} z_t|X_{t-1} \sim & N[-\gamma_2 y_{t-1} + \Pi^{+}_{21}(L)\Delta y_{t-1} \\ & + (I + \gamma_2\delta')z_{t-1} + \Pi^{+}_{22}(L)\Delta z_{t-1}, \Sigma_{22}] \end{aligned} \tag{3.20}$$

Let us now consider that the parameters of interest are given by the cointegrating vector components as well as the corresponding error correction coefficient: $\psi = \text{vec}(\gamma_1, \delta')$. We may now state the following result (see also Johansen, 1992a):

Proposition 1
If the variables are generated according to the cointegrated VAR model and if our parameters of interest are given by $\psi = \text{vec}(\gamma_1, \delta')$, then z_t is weakly exogenous for the parameters of interest ψ if and only if:

$$\gamma_2 = 0$$

Proof.

What has to be proven is that the parameters of interest (γ_1, δ') can be recovered from the conditional model only and that the parameters θ_1 and θ_2 are var iation free in which case a valid sequential cut (in the sense of Engle et al., 1983) is achieved. The parameters of the conditional model (3.18) are: $\theta_1 = \text{vec}(C(1), Q(1), C^{+}_{1}, ..., C^{+}_{n}, Q^{+}_{0}, ..., Q^{+}_{n}, \sigma_{11.2})$. Those of the marginal are denoted by $\theta_2 = \text{vec}(\Pi^{+}_{21,0}, ..., \Pi^{+}_{21,n}, \Pi^{+}_{22,0}, ..., \Pi^{+}_{22,n}, \Pi_{21}(1), \Pi_{22}(1), \Sigma_{22})$. Comparing the expressions (3.14), (3.15) and (3.18) with (3.20) we see that the parameters of the conditional model are not independent of those of the marginal since δ enters both the marginal (through $\Pi_{22}(1)$) and the conditional model (through $Q(1)$). Simply stated, θ_1 and θ_2 are linked via δ and γ_2. In particular, δ enters both θ_1 and θ_2 and γ_2 enters the conditional model through $Q(1)$ and $C(1)$ which are respectively defined by $Q(1) = -\gamma_1\delta' + \sigma_{12}\Sigma_{22}^{-1}\gamma_2\delta'$ and $C(1) = \gamma_1 - \sigma_{12}\Sigma_{22}^{-1}\gamma_2$. The variation freedom condition is thus clearly violated. Consider now the case where $\gamma_2 = 0$. In this case, δ' only enters in the conditional model and the link between θ_1 and θ_2 disappears. From the multivariate normal density we know that $(\Sigma_{22}, \sigma_{12}\Sigma_{22}^{-1}, \sigma_{11.2})$ are variation free so that the parameters θ_1 and θ_2 are variation free and ψ is only a function of θ_1 as seen from the definitions of $Q(1), C(1)$. □

Strong exogeneity will hold if in addition the remaining terms in $\Pi_{21}(L)$ are jointly zero, i.e. if y_t does not Granger-cause z_t.

Discussion

In Proposition 1, the parameters of interest are the long run parameters of the general model (3.12) in which case the only condition is that the conditioning variables are not "error correcting", i.e. that the cointegrating vector does not appear in the equation for the conditioning variables. Also, remember that we have assumed that there are no further restrictions on the VAR. Under these conditions we may validly conduct inference from the single equation (conditional) ECM (3.18) only. This implies that we interpret a single equation ECM as derived by conditioning within a cointegration framework. This is the interpretation used for example by Boswijk (1989) and Johansen (1992a). Since the framework we use is based on a finite order VAR model, weak exogeneity is the sufficient concept to study for efficient inference in conditional models. If we had allowed moving average terms and considered a so-called *triangular ECM*, then the stronger concept of strong exogeneity would be required (see e.g. Phillips, 1991a, Saikkonen, 1991a). Notice however that even in this case, the condition to satisfy for being able to conduct optimal inference in a single equation model boils down to the nullity of a parameter which corresponds, in our set-up, to the weight of the cointegrating vector in the model for the conditioning variables. Following Boswijk (1992a,b) this can be illustrated with the simple bivariate triangular ECM used in the Monte Carlo study of Phillips and Hansen (1990) and Phillips and Loretan (1991).

$$y_t = \alpha z_t + v_{1t} \tag{3.21}$$

$$\Delta z_t = v_{2t} \tag{3.22}$$

where

$$\begin{bmatrix} v_{1t} \\ v_{2t} \end{bmatrix} = \begin{bmatrix} \varepsilon_{1t} \\ \varepsilon_{2t} \end{bmatrix} + \begin{bmatrix} 0.3 & -0.4 \\ \gamma & 0.6 \end{bmatrix} \begin{bmatrix} \varepsilon_{1t-1} \\ \varepsilon_{2t-1} \end{bmatrix} \tag{3.23}$$

and

$$\begin{bmatrix} \varepsilon_{1t} \\ \varepsilon_{2t} \end{bmatrix} \sim N\left[\begin{pmatrix} 0 \\ 0 \end{pmatrix}, \begin{pmatrix} 1 & \rho \\ \rho & 1 \end{pmatrix}\right]$$

As pointed out for example in Phillips (1991a), efficient inference on the cointegrating vector parameter (α) necessitates $\gamma = 0$, i.e. *strong exogeneity* of z_t. The link with Proposition 1 can be seen if we rewrite (see Boswijk, 1992a,b) this triangular ECM in a conditional and a marginal ARMAX model:

$$\begin{aligned} \Delta y_t = -(1.6 + \rho\gamma) \quad (& \quad y_{t-1} - \alpha z_{t-1}) + (\beta + \rho)\Delta z_t \\ - & \quad (0.4 + 0.3\rho)\Delta z_{t-1} + c(L)v_t \end{aligned} \tag{3.24}$$

$$\Delta z_t = \gamma(y_{t-1} - \alpha z_{t-1}) - 0.3\Delta z_{t-1} + c(L)\varepsilon_{2t} \tag{3.25}$$

where $c(L) = 1+0.9L+(0.18+0.4\gamma)L^2$ and $v_t = \varepsilon_{1t} - \rho\varepsilon_{2t}$. We see that γ corresponds to the weight of the cointegrating vector in the marginal equation (3.25). The nullity of this parameter again allows for efficient inference on the cointegrating vector by using the conditional model (3.24) only. In this set-up however, the value of γ determines whether z_t is *strongly exogenous* for the long run parameter α.

If we come back to Proposition 1, it is worth relating the condition for weak exogeneity to the condition of Granger non-causality from y_{t-1} to z_{t-1}. As pointed out, for example, in Engle et al. (1983), Geweke (1984), conditions for Granger non-causality do not usually include any assumption about the parameter so that it must be completed by an assumption of weak exogeneity (which results in strong exogeneity) in order to allow for efficient inference in a conditional model. Conditions for Granger non-causality have recently been studied in the framework of cointegrated VAR by several authors including Mosconi and Giannini (1990) and Phillips and Toda (1991). Within a framework very similar to the cointegrated VAR (3.13), Phillips and Toda (1991), for example, point out that Granger non-causality from y_t to z_t requires $\Pi_{21,i} = 0, \forall i$. This implies by definition that γ_2 should be zero. We thus have a very specific case where Granger non-causality is sufficient (but not necessary) for weak exogeneity of z_t with respect to ψ.

This however does not rule out the possibility that the short run coefficients are a function of the parameters of the marginal model. This is seen from the definition of the parameters of the polynomials $C(L)$ and $Q(L)$ defined in (3.14) and (3.15) which involve $\Pi_{21,i}$ and $\Pi_{22,i}$. So the regression model implied by the conditional expectation has parameters which are themselves a function of the parameters of the marginal process and it is likely that important changes affecting these will imply a loss of invariance of the parameters of the conditional model (3.18). In this case, the definition of super exogeneity (Engle et al., 1983) is not fulfilled if the short run coefficients are parameters of interest, unless further conditions are fullfilled.

Proposition 2
If $\psi = vec(\Pi_{11,0}, .., \Pi_{11,n}, \Pi_{12,0}, .., \Pi_{12,n}, \gamma_1, \delta')$, i.e. if the parameters of interest are both the long run and short run coefficients from the first equation of the cointegrated VAR, then sufficient conditions for z_t to be weakly exogenous for ψ are given by

(i) $\gamma_2 = 0$

(ii) $\sigma_{12} = 0$

This is straightforward from the derivation of the single equation ECM. If $\gamma_2 = 0$ and $\sigma_{12} \neq 0$ then the short run coefficient cannot be recovered from the conditional model since $C(L)$ and $Q(L)$ do depend on the parameters of the marginal model. If $\gamma_2 \neq 0$ and $\sigma_{12} = 0$, single equation estimation will still be less efficient than full information maximum likelihood estimation of the joint model which would explicitly deal with the cross equation restriction and so z_t cannot be weakly exogenous. Such an estimation procedure is proposed by Johansen (1988, 1990a, 1991a).

Since short run dynamic parameters usually give insight into the short run economic rigidity or the deviation from well defined economic hypotheses, they should not be understated in dynamic econometric representations of economic phenomena. A simple example is given in the import demand functions studied in Urbain (1988) where the asymmetry in short run price elasticities is as important as the long run symmetry usually postulated. However, consideration of the short run dynamic coefficients of (3.12) as potential parameters of (theoretical) interest is subject to caution given the usual lack of economic meaning of traditional short run dynamic coefficients in VAR models, unless the factorisation into conditional and marginal models implies super exogeneity. As pointed out by Gourieroux and Monfort (1989, p.462), exogeneity concepts loose much of their sense when they are not discussed in the context of structural models. Notice that other sets of sufficient conditions exists, including the $\{C_i\}$ and $\{Q_i\}$ in (3.14) and (3.15) are structural.

If instead of being the coefficient of (3.12) our parameters of interest are those of the conditional model (3.18) then we have the following corollary:

Corollary 1
If the parameters of interest are those of the conditional model (3.18), given by $\psi' = vec(C(1), C^+(B), Q(1), Q^+(B), \sigma_{11.2})$ *then, by construction,* z_t *is automatically* **predetermined** *(in the sense of Engle et al., 1983) in the regression model implied by the conditional expectation (3.18). If* $\gamma_2 = 0$*, then* z_t *is also weakly exogenous for the parameters of interest.*

This result is straightforward as ζ_t and ε_{2t} are uncorrelated by construction. We have here an interesting example of a case where a variable is predetermined but not weakly exogenous for the parameters of interest unless $\gamma_2 = 0$.

What this brief discussion attempts to point out is that the notion of weak exogeneity is a characteristic of a variable in a model for well defined parameters of interest, and thus not an intrinsic characteristic. Therefore, one should clearly state the framework which is used. This will become even more clear in the next section.

3.3 Error Correction Models in Structural Form

3.3.1 The error correction system in structural form

When the parameters of interest are both long run parameters and the short run parameters, the VAR representation or the error correction model in reduced form do not provide us with a useful parametrisation unless weak exogeneity is satisfied. Further, Δz_t often includes endogenous variables, so we next analyse structural equations.

We suppose that Assumptions 1 and 3 still hold and that the variables are thus jointly generated by a finite order cointegrated VAR model with Gaussian innovations. As pointed out inter alia by Hendry and Richard (1983), in economic theory, interest lies in so-called structural relations rather than in derived equations. A general formulation of the restrictions implied by economic theory can be expressed by means of a set of linear behavioral relationships of the following type (see inter alia Hendry and Richard, 1982, Engle et al., 1983):

$$A_0\mu_t + \sum_{i=1}^{n} A_i x_{t-i} = 0 \tag{3.26}$$

where μ_t is defined as the conditional expectation of $x_t|X_{t-1}$ and $A = (A_0, A_1, \ldots, A_n)$ may be subject to normalisation and sufficient a priori identifying restrictions. This implies that $A_0\Pi_i + A_i = 0$ for $i = 1, .., n$. Assuming A_0 as a nonsingular $(k \times k)$ matrix and premultiplying the derived VAR model by A_0 we have:

$$A_0 x_t = A_0[\sum_{i=0}^{n} \Pi_i x_{t-i}] + A_0\varepsilon_t \tag{3.27}$$

which can be written as a traditional Autoregressive Distributed Lag model (ADL):

$$A(L)x_t = v_t \tag{3.28}$$

where $A(L) = A_0\Pi^*(L)$ and $v_t = A_0\varepsilon_t$. The derivation of the *error correction system in structural form* is obtained by writing $A(L) = A(1)L + (1 - L)A^+(L)$:

$$A^+(L)(1-L)x_t = -A(1)x_{t-1} + v_t \tag{3.29}$$

The major difference from the ECM in reduced form is that here $A^+(0) \neq I_k$ and the disturbance covariance matrix is now $A_0\Sigma A_0' = \Omega$. If the identifying restrictions on A are valid, then (3.29) will be congruent with the data evidence (see Monfort and Rabemanajara, 1990, Hendry and Mizon, 1990).

3.3.2 Single equation error correction model in structural form and weak exogeneity

We now partition $x_t' = (y_t, z_t')$ where y_t is a scalar time series and z_t is assumed to be a $(k-1) \times 1$ vector of time series. We may rewrite (3.29) accordingly as:

$$\begin{bmatrix} A_{11,0} & A_{12,0} \\ A_{21,0} & A_{22,0} \end{bmatrix} \begin{bmatrix} \Delta y_t \\ \Delta z_t \end{bmatrix} = \begin{bmatrix} A_{11}^+(L) & A_{12}^+(L) \\ A_{21}^+(L) & A_{22}^+(L) \end{bmatrix} \begin{bmatrix} \Delta y_{t-1} \\ \Delta z_{t-1} \end{bmatrix}$$
$$- \begin{bmatrix} A_{11}(1) & A_{12}(1) \\ A_{21}(1) & A_{22}(1) \end{bmatrix} \begin{bmatrix} y_{t-1} \\ z_{t-1} \end{bmatrix} + \begin{bmatrix} v_{1t} \\ v_{2t} \end{bmatrix} \tag{3.30}$$

where $A_{ij}^{+}(L) = -\sum_{k=1}^{n-1} A_{ij,k}^{+} L^k$

with $A_{ij,k}^{+} = -\sum_{m=k+1}^{n} A_{ij,m}$,

$A(1) = A_0\Pi(1) = A_0\gamma\alpha'$

$A_{11}(1) = A_{11,0}\gamma_1 + A_{12,0}\gamma_2$

$A_{12}(1) = -A_{11,0}\gamma_1\delta' - A_{12,0}\gamma_2\delta'$

$A_{21}(1) = A_{21,0}\gamma_1 + A_{22,0}\gamma_2$

$A_{22}(1) = -A_{21,0}\gamma_1\delta' - A_{22,0}\gamma_2\delta'$

Our interest will lie in the first equation of the system (3.30) which we shall call a **single equation error correction model in structural form:**

$$
\begin{aligned}
A_{11,0}\Delta y_t + A_{12,0}\Delta z_t = & A_{11}^{+}(L)\Delta y_{t-1} + A_{12}^{+}(L)\Delta z_{t-1} \\
& -A_{11}(1)y_{t-1} - A_{12}(1)z_{t-1} + v_{1t}
\end{aligned}
\tag{3.31}
$$

If we rearrange the terms and premultiply[8] by $A_{11,0}^{-1}$ we get the reduced form of (3.31) whose parameters will be denoted by θ_1'

$$
\begin{aligned}
\Delta y_t = & -A_{11,0}^{-1}A_{12,0}\Delta z_t + A_{11,0}^{-1}A_{11}^{+}(L)\Delta y_{t-1} + A_{11,0}^{-1}A_{12}^{+}(L)\Delta z_{t-1} \\
& -A_{11,0}^{-1}A_{11}(1)y_{t-1} - A_{11,0}^{-1}A_{12}(1)z_{t-1} + v_{1t}^{+}
\end{aligned}
\tag{3.32}
$$

where

$v_{1t}^{+} = \varepsilon_{1t} + A_{11,0}^{-1}A_{12,0}\varepsilon_{2t}$,

$\mathrm{Var}(v_{1t}^{+}) = \omega_{11} = \sigma_{11} + A_{11,0}^{-1}\sigma_{12}A_{12,0}' + A_{11,0}^{-1}A_{12,0}\sigma_{21} + A_{11,0}^{-1}A_{12,0}\Sigma_{22}$

We may now derive a set of sufficient conditions for the weak exogeneity of z_t for the parameters of interest. Let us first denote these as the short run

[8]It is clear that if, as is usually done, we normalise A_0 such that it has unit elements on the diagonal, then $A_{11,0}^{-1} = 1$ and can be disregarded in the analysis. We however do not impose this in order to make the generalisation to conditional ECM with more than one component in y_t straightforward.

parameters, the components of the cointegrating vector (and its coefficient) of the first equation of the structural system:

$$\psi' = \text{vec}(A_{12,0}, A_{11,0}^{+}, ..., A_{11,n}^{+}, A_{12,0}^{+}, .., A_{12,n}^{+}, \gamma_1, \delta', \omega_{11})$$

Proposition 3
Assume that the variables are jointly generated according to the cointegrated VAR model. A set of sufficient conditions for the weak exogeneity of z_t for ψ' in the structural ECM (3.31) is given by

(i) $A_{11,0}\sigma_{12}\Sigma_{22}^{-1} + A_{12,0} = 0$

(ii) $A_{21,0} = 0$

(iii) $\mathit{rank}(A_{22,0}) = k - 1$

(iv) $\gamma_2 = 0$

(v) ψ' *is only a function of* θ_1'

Proof.

The proof comes directly from Richard (1980, Theorem 3.1). Conditions (i)-(ii) are overidentifying restrictions. If they hold, then replacing $A_{11,0}^{-1}A_{12,0} = -\sigma_{12}\Sigma_{22}^{-1}$ in (3.31), we see that (3.32) corresponds to (3.18) if in addition (iii) holds. In the preceding section we derived the regression model implied by the conditional expectation as (3.18) and found that its parameters, under (iv), are variation free with those of the marginal model. □

We can also, in the spirit of limited information analysis, complete the first structural equation by a set of unrestricted reduced forms, in which case the following result holds:

Proposition 4
If we replace the remaining relations in (3.30) by a set of unrestricted reduced form equations then sufficient conditions are given by

(i') $\omega_{12} = 0$

(ii') ψ' *are the parameters of the first "behavioral" relations*

(iii') $\gamma_2 = 0$

Proof.

If we have unrestricted reduced forms then (ii) and (iii) hold by construction. Moreover, under these conditions, (i') and (i) are equivalent since

$$\omega_{12} = (A_{11,0}\ A_{12,0}) \begin{pmatrix} \sigma_{11} & \sigma_{12} \\ \sigma_{21} & \Sigma_{22} \end{pmatrix} \begin{pmatrix} A'_{21,0} \\ A'_{22,0} \end{pmatrix}$$

□

In contrast to the previous section, when the short run dynamic is of interest for the modeller (and when some identifying restrictions on the short run dynamic are imposed) and hence if the short run parameters are parameters of ("theoretical") interest, then the condition derived in Proposition 1 (see also Johansen, 1992a) that the cointegrating vector does not appear in the equations of the conditioning variables is no longer sufficient for valid inference from single equation ECM and more generally from a conditional sub-model.

3.4 Inference on Weak Exogeneity in ECMs

3.4.1 Are orthogonality tests useful in single equation ECMs

Whether orthogonality tests (for surveys in a classical framework, see Holly, 1985, Pesaran and Smith, 1990) are useful in the context of single equation error correction models will depend upon the interpretation of the latter.[9]

[9]A further problem with orthogonality tests in this set-up is the possibility of a non-standard distribution as we will see in the next chapter.

Consider briefly that single equation ECMs derive from conditioning in a general cointegration framework as presented above. To simplify the discussion, we rule out the problem of the existence of non-linear cross-equation restrictions. Testing for the weak exogeneity of, say, the second variable (i.e. the second component of our x_t vector process) by means of an orthogonality test means that we would conduct the analysis on the first equation of the system and check the null hypothesis of zero residual covariance. In practical applications (see Steel, 1987a, Urbain, 1988, Lubrano et al., 1986, ...), single equation ECMs are specified with contemporaneous potential endogenous variables on the r.h.s., and thus they are similar to a conditional single equation error correction model such as (3.18). The latter has the property (from the multivariate normal distribution) that its derived innovations are uncorrelated with those of the remaining equations in (3.12). It is thus likely that orthogonality tests which seek to detect a non-zero covariance between these will not reject weak exogeneity.

On the other hand, if the single equation error correction model is in structural form, testing for zero covariance between the residuals from (3.31) and ε_{2t} will be a sufficient condition if the parameters of interest are only a function of those of the conditional model and if the overidentifying restrictions are assumed to hold or if the structural model is completed by unrestricted reduced form models. Thus, testing for zero covariance as a test for weak exogeneity is conditional on $\gamma_2 = 0$.

This calls for some clarification about how single equation error correction models are interpreted. Note that within an incomplete modelling framework where we do not allow for predetermined variables, it is not clear whether (3.18) and (3.31) can really be distinguished unless we follow the arguments in Spanos (1990a) where identification of a structural model is viewed as a statistical reparametrisation of the conditional expectation. The natural way to proceed is then to start from an unrestricted VAR (or system ECM in reduced form) model which is then sequentially reduced to a lower dimensional (possibly single equation) structural model. In the context of stationary variables, such a reduction sequence is proposed by Monfort and Rabemanajara (1990). In the case of $I(1)$ variables, a complete set of reductions is also proposed in Hendry and Mizon (1990).

3.4.2 Testing for the presence of cointegrating vectors in the marginal models

As the presence of the error correction term in the various equations is a property of the system-ECM representation (3.12), it appears most natural to check for this hypothesis within such a full model. This can be conducted within the full information maximum likelihood method developed by Johansen (1988, 1990, 1991a) and Johansen and Juselius (1990). Consider the vector autoregressive error correction model:[10]

$$\Delta x_t = \sum_{i=1}^{n-1} \Gamma_i \Delta x_{t-i} + \Gamma_n x_{t-n} + \varepsilon_t \tag{3.33}$$

where $\Gamma_i = -I + \Pi_1 + ... + \Pi_i$ with $i = 1, \ldots, n$. Γ_i are interim multipliers. If $\text{rank}(\Gamma_n) = r < k$ then it may be written as $\gamma\alpha'$ where γ and α are $(k \times r)$ matrices. The rows of α form the cointegrating vectors while the matrix γ contains the factor loadings, i.e. γ gives the weights of the cointegrating vectors in the various equations. Testing for restrictions on this γ matrix will provide us with a test of weak exogeneity if our parameters of interest are the long run parameters only. Note that even if we are interested in the short run coefficients, this procedure may be sufficient for the *rejection* of weak exogeneity. This can easily be conducted using a LR test distributed as $\chi^2(p)$ under the null hypothesis.

However, if we want to avoid the modelling and estimation of the full system one may proceed as follows (see Johansen, 1992a). First perform a cointegration analysis of the partial conditional model to obtain consistent estimates of the cointegrating vectors, and then a regression analysis of the (possibly large) marginal model along with a misspecification check for the presence of the cointegrating vectors.

A similar *LM* test for weak exogeneity has recently been proposed by Boswijk (1991b) in a structural ECM framework.

Note that within their two-step framework, Engle and Granger (1987, p. 273) argue that a simple way to check the weak exogeneity of z_t for the

[10] We have omitted the possible constant term as well as seasonal dummies although their effect on the limiting behaviour of the various test statistics is well documented (inter alia Johansen and Juselius, 1990).

parameters of interest is to estimate an ECM for z_t and test the significance of the error correction term using a traditional t-test.

3.5 Empirical Illustration

In this section we now illustrate the above mentioned points by reanalyzing some empirical studies which investigate weak exogeneity within single equation error correction models. The choice of these studies is also motivated by the fact that they were all derived within a general to specific approach. As the presence of a cross-equation restriction is sufficient for the rejection of weak exogeneity in both types of single equation ECMs, we shall focus on this hypothesis.

3.5.1 Belgium consumption function: Steel (1987)

We first consider the study by Steel (1987a) who considers an annual consumption model for the Belgian economy. His basic hypothesis was the weak exogeneity of disposable income for the parameters of interest. Note that these are considered as both the long run and the short run coefficients. From a general to simple approach, the following single equation error correction model was derived (for complete results, see Steel 1987a), with asymptotic standard errors in brackets

$$\begin{array}{rllll} \Delta co_t = & \underset{(0.05)}{0.33\,\Delta ia_t^*} & - \underset{(0.05)}{0.20\,(co - ia*)_{t-2}} & - & \underset{(0.0004)}{0.0058\,CORR_{t-1}} \\ & & + \underset{(0.0011)}{0.0041 DR_{t-2}} & - & \underset{(0.0004)}{0.0015\,l_{t-1}} \end{array}$$

where co denotes consumption of nondurable goods (excluding services), ia^* personal gross disposable income adjusted for inflation,[11] $CORR$ a correction variable for inflation adjustment, DR the discount rate of the national bank and l accumulated saving with L_0=0 and without inflation. Detailed definitions and constructions of these variables are given in Steel (1987a). In

[11] with the additional characteristic that this variable was calculated under the assumption of zero accumulated saving at $t = 0$, i.e. 1953.

order to test for the weak exogeneity of ia^*, a reaction function (marginal model) has also been estimated:

$$\begin{array}{llllll} \Delta ia_t^* = & -3.09 + & 0.83\, gnp_t + & 0.089\, s_t - & 0.16\, s_{t-1} - & 1.83\, \Delta P_t \\ & (0.36) & (0.10) & (0.02) & (0.01) & (0.10) \\ & -0.84\, l_t + & 0.67\, l_{t-1} + & 0.54\, tco_t - & 0.90\, tco_{t-1} & \\ & (0.11) & (0.09) & (0.18) & (0.13) & \end{array}$$

$R^2 = 0.98$, DW $= 2.11$, sample: 1956-1982, where gnp denotes gross national product, s national savings, P a two-year inflation measure, tco total consumption.

A first simple check is to introduce the error correction term of $(co - ia^*)_{t-2}$ into the reaction function and then to reestimate it. We were unable to reproduce the original results published by Steel (1987a) - see Model A, Table 3.1- . The reason might be that the data we use for total consumption and savings are not the original data used by Steel (1987a).

In Table 3.1, heteroscedastic consistent standard errors are reported in brackets. Other notation include: η_1 which is the Chow test for predictive failure distributed F(4,17) under the null; η_2 is Hendry's predictive failure test, distributed $\chi^2(3)$ under the null of no predictive failure, η_3 is the Lagrange Multiplier test for 3rd order autocorrelation in F-form distributed F(3,18); η_4 is the LM test for first order autoregressive conditional heteroscedasticity (ARCH) distributed $\chi^2(1)$ under the null; while η_5 is the second order RESET distributed F(2,19) under the null.

The introduction of the error correction term appears significant although its coefficient is rather low (see Model B). However the t-test is significant at a 1% level and one can thus conclude that Δia^* may not be considered as weakly exogenous for the long run parameters. If we do not want to rule out the possibility of Granger-causal priority from disposable income to consumption a priori we may extend the reaction function into an ECM form where we thus add lagged Δco_t and the error correction term (Model C and D). This is not in contradiction with the General Instrumental Variables approach used in Steel (1987a) since it allows the instruments to be selected upon their own merits. The ECM term is significant[12] as well

[12] Conventional distributional results are applicable for the t-test statistic since the ECM term imposing unit long run elasticity is stationary (DW: 1.18; ADF = - 4.32). Note

Dep.var.:Δia_t^*	Model A	Model B	Model C	Model D
Constant	-6.860(2.05)	- 6.092(1.51)	-8.251(2.28)	-9.018(1.67)
gnp_t	0.610(0.28)	0.644(0.17)	0.714(0.22)	0.857(0.19)
s_t	-0.035(0.005)	- 0.032(0.04)	- 0.064(0.46)	-0.081(0.04)
s_{t-1}	- 0.183(0.09)	-0.124(0.70)	- 0.194(0.09)	-0.154(0.05)
ΔP_t	-2.354(0.40)	- 2.571(0.27)	-2.409(0.32)	-2.465(0.22)
l_t	-0.041(0.06)	- 0.765(0.45)	-0.071(0.52)	-0.410(0.26)
l_{t-1}	0.043(0.44)	0.681(0.33)	0.109(0.42)	0.397(0.29)
tco_t	0.333(0.35)	0.335(0.23)	0.323(0.23)	0.214(0.22)
tco_{t-1}	-0.335(0.29)	-0.361(0.22)	-0.365(0.21)	- 0.262(0.19)
Δco_{t-1}	-	-	-0.815(0.31)	- 0.564(0.21)
Δco_{t-1}	-	-	0.004(0.40)	-
$(co - ia^*)_{t-2}$	-	0.0736(0.03)	0.047(0.02)	0.049(0.02)
R^2	0.88	0.95	0.93	0.94
DW	2.67	2.40	1.96	2.05
η_1	1.53	1.31	1.32	1.51
η_2	2.48	2.44	2.42	2.64
η_3	1.65	0.92	0.69	0.86
η_4	1.06	1.07	0.71	0.63
η_5	1.14	0.82	0.55	0.70
t-test on ECM	-	2.405	2.361	2.14

Table 3.1: Auxiliary regression for Δia_t^*

	$\hat{v}_A$	$\hat{v}_B$	$\hat{v}_C$	$\hat{v}_D$
coefficient	-0.0809	-0.1814	- 0.1886	0.4844
standard error	0.207	0.228	0.228	0.167
t-test	-0.390	-0.797	-0.827	2.893

Table 3.2: Exogeneity tests by variable addition

as the first lagged Δco_t. At the bottom of the table are the test statistics which we used and the last row are the individual t-tests for the significance of the error correction term. Δia^* can thus not be treated as weakly exogenous for the long run parameters of interest, namely the long run (unit) elasticity. Moreover, these results also imply that disposable income does Granger-cause consumption.

Alternatively, using traditional limited information exogeneity tests we may also check for the weak exogeneity of Δia^* for the parameters of interest (which are supposed to be both the long and short run coefficients) by adding the estimated residual obtained from the reaction function to the original error correction formulation. We here use a variable addition type exogeneity test which is based on the reverse likelihood factorisation of the joint distribution compared to that used by Steel (1987a).[13] The results obtained are straightforward to interpret since this additional term appears insignificant (Table 3.2) according to the individual t statistic.

In Table 3.2, $\hat{v}_i$ is the residual obtained by fitting the marginal model i from Table 3.1. The only case where weak exogeneity is rejected is when the auxiliary model for Δia_t^* is model D with the ECM term and the lagged Δco_t. This also shows that the outcome of these procedures is extremely

that Johansen's full information maximum likelihood approach could not be implemented given the small size of the sample. Building a VAR for 5 variables and at least one lag leaves a very small of degrees of freedom available.

[13] We do not know of any studies investigating the small samples properties of these test statistics and thus we do not know of their relative merits in such a small sample of 30 annual data points.

sensitive to the choice of variables (instruments) that are used. For the three other auxiliary models, the results reinforce Steel (1987a)'s findings but also point out that orthogonality tests may be of little use when cross equation restrictions are present.

3.5.2 Aggregate import demand function: Urbain (1988)

In a previous empirical study (Urbain, 1988) we analysed aggregate import demand functions for some OECD countries including Belgium, The Netherlands, UK and USA. The hypothesis tested was whether we could treat import prices as weakly exogenous for the parameters of interest. We will only report the results for the UK but similar results are obtained for the other countries. The variables used in the analysis are those of the traditional imperfect substitutes models namely volume of imports (M), real GDP (Y), import price (PM), wholesale (domestic) price index (PD) and relative prices defined as $PR = PM/PD$. The data are annual figures covering the period 1952-1986. They are all taken from the IMF financial statistics. Engle and Granger's two step method and a careful dynamic specification of the short run dynamic led to the following selected conditional error correction model where all variables are expressed in logs (for complete detailed results, see Urbain, 1988):

$$\begin{array}{llll} \Delta m_t = \underset{(0.02)}{0.063} + & \underset{(0.25)}{1.043}\ \Delta y_t + & \underset{(0.27)}{0.386}\ \Delta pd_t - & \underset{(0.09)}{0.298}\ \Delta pm_t \\ \quad - \underset{(0.08)}{0.106}\ \Delta pm_{t-2} & + \underset{(0.06)}{0.441}\ \Delta_2 pr_t - & \underset{(0.12)}{0.572}\ \Delta_2 m_{t-1} - & \underset{(0.23)}{0.919}\ EC_{t-3} \end{array}$$

where EC is the residual from the cointegrating regression which gave the following long run relationship (DW=1.250, ADF=-3.364, DF=- 4.24):

$$m = 3.06 + 1.255\ y + 0.40\ (pm - pd)$$

A reaction function for Δpm_t was also estimated:[14]

$$\begin{aligned}\Delta pm_t = \underset{(0.03)}{-0.86} + \underset{(0.43)}{1.477}\,\Delta y_t + \underset{(0.16)}{1.345}\,\Delta pd_t - \underset{(0.17)}{0.372}\,\Delta ex_t \\ - \underset{(0.15)}{0.323}\,\Delta ex_{t-1} + \underset{(0.05)}{0.082}\,\Delta pw_t - \underset{(0.03)}{0.066}\,\Delta pw_{t-1}\end{aligned}$$

where pw is the log of world consumption price expressed in growth rates used as a proxy of world inflation and ex is the current exchange rate (national currency/DTS). Adding the cointegrating vector lagged one period gave:

$$\begin{aligned}\Delta pm_t = \underset{(0.19)}{-0.93} + \underset{(0.61)}{1.315}\,\Delta y_t + \underset{(0.28)}{1.003}\,\Delta pd_t - \underset{(0.21)}{0.514}\,\Delta ex_t \\ - \underset{(0.11)}{0.291}\,\Delta ex_{t-1} + \underset{(0.07)}{0.101}\,\Delta pw_t - \underset{(0.02)}{0.029}\,\Delta pw_{t-1} + \underset{(0.06)}{0.182}\,EC_{t-1}\end{aligned}$$

The individual t-statistic on the coefficient of the error correction term in this auxiliary model gives a value of 3.035 and we may thus not treat import prices as weakly exogenous for the parameters of interest within a single equation import demand function.

3.5.3 The UK demand for money: Lubrano et al. (1986), Steel and Richard (1991)

The last empirical exercise we conduct is on the UK money demand studied by Lubrano et al.(1986) - hereafter LPR- and reanalyzed recently by Steel and Richard (1991) - hereafter SR. The initial interest of LPR was in the effect of the introduction of measures known as Competition and Credit Control (CCC) that occurred in the UK in 1971. As the interest rate setting process was supposed to be fundamentally modified with the introduction of CCC, interest rates might well have changed exogeneity status over the period before and after the introduction of CCC. Therefore LPR analysed the exogeneity status of interest rates for the parameters of a single equation

---

[14] In fact two different auxiliary models were estimated, both leading to the non rejection of weak exogeneity. We only report the auxiliary model which appeared to be the most reliable.

demand for money in "structural ECM form" and found - within a Bayesian framework - that (*i*) weak exogeneity could not be rejected for the post-CCC regimes; (*ii*) money does not Granger cause interest rates in the pre-CCC regime. SR extended LPR analysis by investigating the joint exogeneity of prices and interest rates which could also not be rejected over an extended sample covering the period 1955(ii)-1986(ii).

Both studies use Bayesian techniques for testing for weak exogeneity which amounts to testing for zero covariance between their conditional (ECM) and a marginal (reaction function) model without explicitly considering the cross-equation restriction implied by the ECM formulation.[15] We shall not go into the details of interpretation of the results obtained, or the chosen variables in these studies but shall instead reanalyze these data by using the full information analysis proposed by Johansen (1988, 1991a) and Johansen and Juselius (1990). In the line of LPR and SR we consider two sub-periods denoted respectively by periods A (1955(ii)-1971(iii)) and B (1971(iv)- 1986(ii)). The variables used are: M for the M3 personal sector monetary aggregates, Y for real personal disposable income, P for the deflator of Y, R for the local authorities short term interest rates. For detailed definitions and data sources we refer to appendix A in SR. The data are quarterly unadjusted figures. Our notation will be as far as possible compatible with that of LPR and SR: Δ for first difference, L for natural log.

The underlying long run (steady state) solution for both money demand equations assumes constant velocity of circulation of money and thus imposes in the long run that the coefficients of money, prices and income are equal with opposite sign while the interest rates effect is left unrestricted. Note that SR use ΔLM as dependent variable and ΔLP as explanatory variable, while LPR use $\Delta LMP = \Delta(LM_t - LP_t)$.

[15] Within a Bayesian framework the variation free condition of weak exogeneity is replaced by a condition of *prior independence* (see Steel and Richard, 1991). This makes a difference when we are at the inference level but does not rule out cross equation restrictions which apply at the level of the model building.

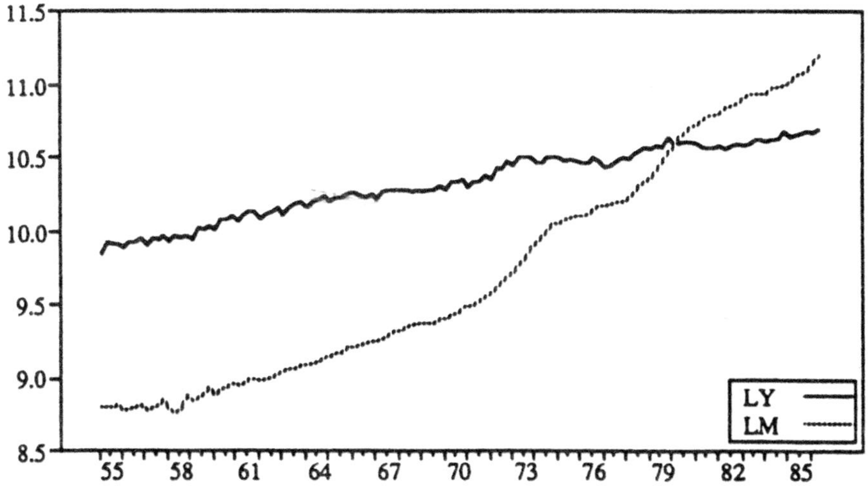

Figure 3.1: Real disposable income and M3 (in logs)

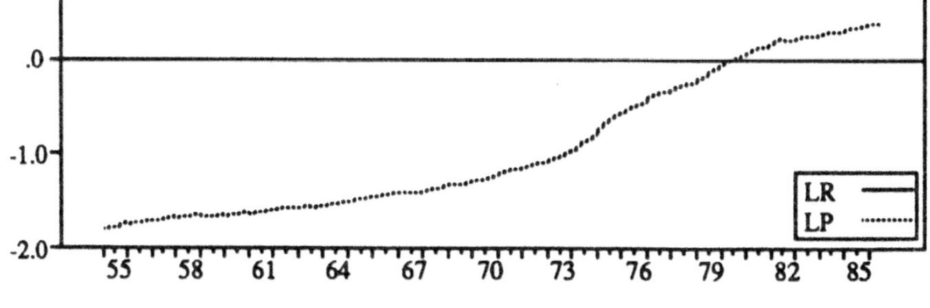

Figure 3.2: Deflator of the real disposable income and short term interest rates (in logs)

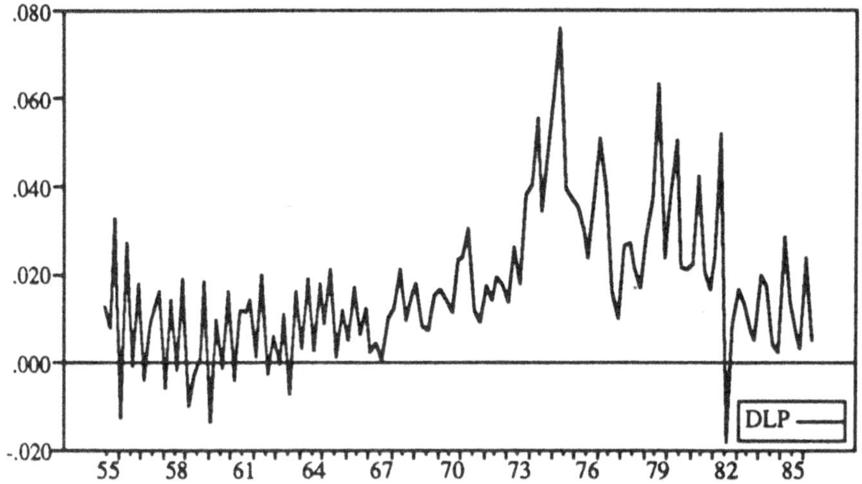

Figure 3.3: Deflator of the real disposable income (in first diff.)

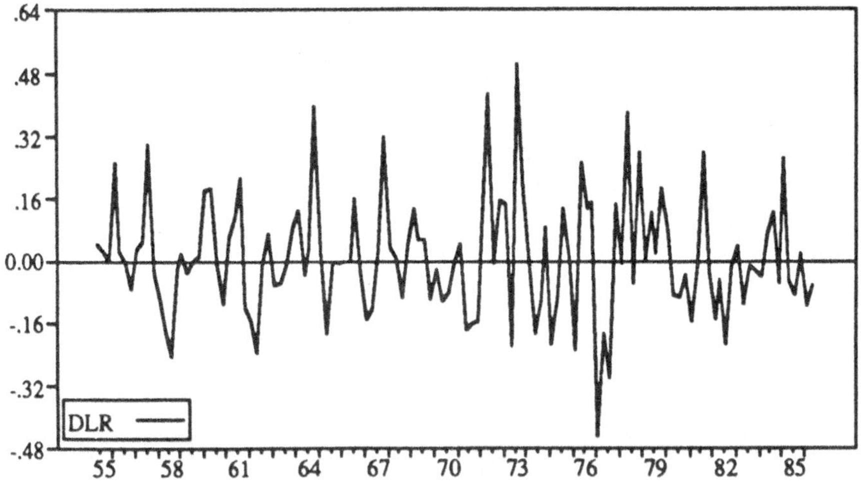

Figure 3.4: Short run interest rates (in first diff.)

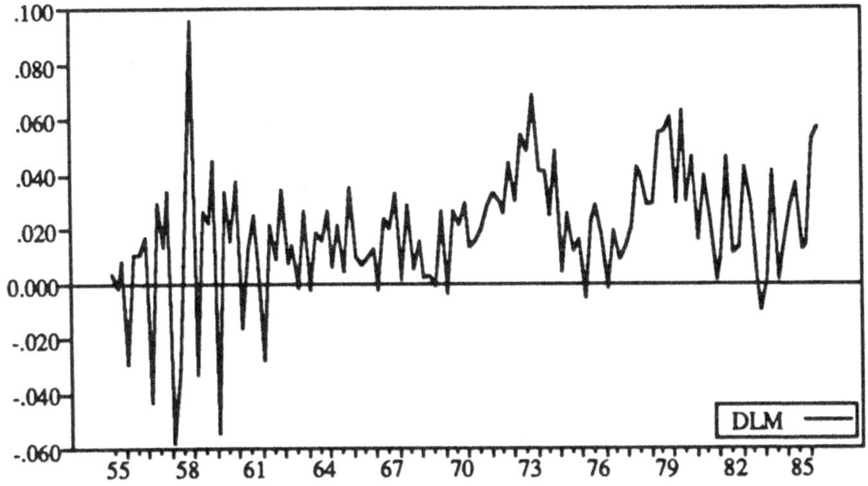

Figure 3.5: M3 (in first diff.)

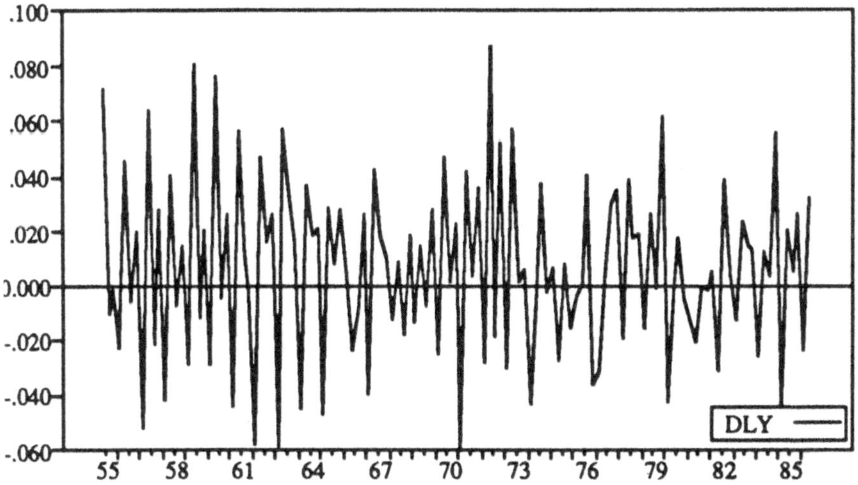

Figure 3.6: Real personal disposable income (in first diff.)

Here, we shall fit a VAR model to $x_t = (LM_t, LY_t, LP_t, LR_t)'$. Contrary to related empirical studies on UK money demand it seems unnecessary to impose price homogeneity to have $I(1)$ processes at least when the subperiods are separately considered. As the variables are likely to possess some linear trend components (see figures) we fitted the following model to the data:

$$\Delta x_t = \sum_{i=1}^{n-1} \Gamma_i \Delta x_{t-i} + \Gamma_n x_{t-n} + \mu + \Phi D_i + \varepsilon_t \tag{3.34}$$

where μ is a constant and D_i a set of centered seasonal dummies. The following hypotheses were tested within Johansen's approach:

- H_1: the number of cointegrating relations: the rank of $\Gamma_n = \gamma\alpha'$
- H_2: restrictions on the cointegrating vector(s) α
- H_3: restriction on the loading matrix γ

H_1 is simply a test to validate some long run relations among the variables. H_2 will test restrictions on the long run relations (cointegrating vectors). Here, we shall be interested in the restrictions implied by the hypothesis of constant velocity of circulation of money imposed by LPR and SR. Finally H_3 enables us to test restrictions on the loading matrix, i.e. it enables us to test whether cointegrating vectors enter a particular equation or not. This is essential for any assessment of weak exogeneity.

To implement Johansen's approach[16] for the data set a first decision must be made on the order of the model. Given the results in SR we suspect the optimal lag length to differ across the two regimes, the post CCC results giving some indication that one might need a higher dimensional VAR model. On the basis of the Box-Pierce Q statistic and the Jarque and Bera (1980) normality test we decided to keep $n = 3$ during period A and $n = 5$ during the post-CCC period. Note that SR derived their specification by starting from an unrestricted ADL with 5 lags.[17] Complete results are reported in Tables 3.3-3.4.

[16] The computations were carried with RATS version 3:10.

[17] Performing the same exercise on the entire sample led us to select n=4.

Period A : 1955:02-1971:03				Period B : 1971:04-1986:02			
			Eigenvalues				
0.4081	0.1987	0.1205	0.0273	0.5162	0.2617	0.0782	0.0179
			H_1 :Testing the number of cointegrating vectors				
λ_{max}	Trace test	95% λ_{max}			95% Trace	λ_{max}	Trace test
1.64	1.64	8.08	r≤3		8.08	1.011	1.01
7.58	9.21	14.60	r≤2		17.84	4.56	5.57
13.08	22.29	21.28	r≤1		31.26	16.	22.523
30.94	53.23	27.34	r=0		48.42	40.66	63.183
			Normalised Cointegrating vector				
δ = (1, -0.517, -1.133, -0.020)				δ = (1, -3.315, -0.286, -0.346)			
			Factor Loading				
γ = (-0.441, -0.069, 0.123, 2.360)				γ =(-0.109, 0.073, -0.071, -0.428)			
			$H_2 : \delta_1 = -\delta_2 = -\delta_3$				
	13.051		$\chi^2(2)$			19.215	
			$H_3 : \gamma_3 = \gamma_4 = 0$				
	15.785		$\chi^2(2)$			17.852	

Table 3.3: Money demand in the UK: MLE approach with $n = 3$

For both sub-periods the likelihood ratio test for the number of cointegrating relations favors a unique cointegrating vector.[18] We normalised the first component of this cointegrating vector so that money has a unit coefficient.

Although it is not the purpose of this work to investigate the implications of such important differences in the long run estimates it is surprising to note the important shift occurring in the estimated coefficients. Note that these differences remain valid with all the different lag lengths we have tried in the empirical analysis. The restrictions implied by the constant velocity of money hypothesis are tested by means of a likelihood ratio test asymptotically distributed, under the null, as a $\chi^2(2)$. For both subperiods, this hypothesis is rejected at a 1 % level.

The final hypothesis concerns the loading vector γ which distributes the cointegrating relations in the various equations. The unrestricted estimates are given by: (-0.441, -0.063, -0.123, 2.360) and (-0.066, 0.156, -0.085, 0.102)

[18] For the second subperiod, we should accept the presence of a second cointegrating relation. However, its graphical representation displays some nonstationary behaviour so that we only retained a unique cointegrating relation. Moreover, using full sample results, one cannot reject the hypothesis of a unique cointegrating relation.

Period A : 1955:02-1971:03				Period B : 1971:04-1986:02			
			Eigenvalues				
0.2499	0.1668	0.1297	0.0072	0.4705	0.3499	0.1511	0.0109
			H_1 :Testing the number of cointegrating vectors				
λ_{max}	Trace test	95% λ_{max}			95% Trace	λ_{max}	Trace test
3.22	3.22	8.08	r$\leq$3		8.08	0.5912	0.5912
7.49	10.72	14.60	r$\leq$2		17.84	8.85	9.94
12.94	23.16	21.28	r$\leq$1		31.26	23.26	32.70
24.87	48.03	27.34	r=0		48.42	34.28	66.99
			Normalised Cointegrating vector				
$\delta = (1, -0.526, -1.065, 0.028)$					$\delta = (1, -2.880, -0.262, -0.650)$		
			Factor Loading				
$\gamma = (-0.505, -0.249, 0.012, 5.263)$					$\gamma =(-0.066, 0.156, -0.086, 0.102)$		
			$H_2 : \delta_1 = -\delta_2 = -\delta_3$				
	11.909		$\chi^2(2)$			8.390	
			$H_3 : \gamma_3 = \gamma_4 = 0$				
	8.107		$\chi^2(2)$			10.654	

Table 3.4: Money demand in the UK: MLE approach with $n = 5$

for period A and B respectively. Testing $H_0 : \gamma_3 = \gamma_4 = 0$ is a proper test for the weak exogeneity of interest rates and prices for the long run parameters. This procedure is also sufficient to reject weak exogeneity even if the parameters of interest are both the long and short run coefficients. The likelihood ratio test for this hypothesis is distributed as a $\chi^2(2)$ under the null. For the first period we get a value of 15.785 while for period B we obtain a value of 10.654. In both cases the null hypothesis that the cointegrating relations are not present in the equation determining interest rates and prices is rejected at a 5 % level. Neither the price nor the interest rate variables may thus be treated as weakly exogenous for the parameters of interest. Note the important differences in the estimated factor loadings between the two periods. Similar results are reported in SR and to a lesser extend by LPR, but remember that the latter covers only the period 1955(ii)-1971(iii) for period A and 1971(iv)-1981(ii) for period B, so that the sample size is small in both regimes. Also, excluding a trend imposes the restriction that LY_t is not stationary around a linear trend

3.6 Conclusions

The literature which investigates the concept of weak exogeneity in error correction models has been confined to the case where the parameters of interest were primary the long run parameters, i.e. the cointegrating vectors (see Johansen, 1992a, Engle and Yoo, 1991). ECMs may however have different types of interpretation and this may be misleading as particular variables may be weakly exogenous for these, but not for the short run parameters. Weak exogeneity tests, such as those proposed by Johansen (1992a) and Boswijk (1991b), should be extended to allow testing exogeneity when short run parameters are also parameters of interest.

On the other hand, most of the traditional orthogonality tests as surveyed by Pesaran and Smith (1990) and Holly (1985) are confined to the possible detection of non zero covariance. Alone, they are thus not sufficient to address this problem within ECMs derived from a cointegration framework since a variable which appears to be weakly exogenous according to the test statistic may not be weakly exogenous for the long run parameters of interest if it is error correcting. [19]

The review of some previous empirical work has illustrated that when one works within an incomplete model (i.e. a single structural equation completed with auxiliary models) it is very unlikely that the cross-equation restriction will be taken into account.

[19] As pointed out by Dolado (1992), weak exogeneity is not always necessary to sustain inference on the long run parameters in conditional models since the diagonality of the long run covariance matrix may sometimes suffice, even if the conditioning variable is error correcting, i.e. not weakly exogenous. In most cases however, the absence of the cointegrating vector in the model for the conditioning variable is a sufficient and *necessary* condition for weak exogeneity w.r.t. the long run parameters (see Hendry (1992) for the pivotal role of the weak exogeneity condition of Proposition 1 in validating inference in conditional cointegrated models).

Chapter 4

Testing for Weak Exogeneity in Error Correction Models

4.1 Introduction

In the previous chapter we have mentioned how one could investigate the presence or absence of a cointegrating vector in the generating model for the conditioning variables. In particular, Johansen (1992a) and Boswijk (1991b) have proposed to test the hypothesis of weak exogeneity w.r.t. the long run parameters but do not consider any orthogonality conditions since the fact that an explanatory variable is error correcting is sufficient to reject weak exogeneity for the long run parameters. If we are in a structural framework, this condition is however not always sufficient when we are theoretically interested in the parameters of the short run dynamic in which case an additional orthogonality condition has to be verified. Although orthogonality tests will not take into account the possible violation of the variation free condition of weak exogeneity (see Engle et al., 1983), they may remain useful if the model used in empirical analysis is a structural error correction model, if the parameters of interest include both short and long run coefficients and if the cointegrating vector is known to be absent from the marginal models.

However, the presence of integrated processes in econometric models is also known to induce distributional problems for many traditional test statistics (see inter alia Park and Phillips, 1988). Since orthogonality tests are often used in empirical work to check for weak exogeneity or for general misspecification, it may be useful to investigate their behaviour in the presence of (co)integrated variables.

In the stationary framework, testing for exogeneity is usually conducted in a limited information or instrumental variables framework (Holly, 1985; Engle, 1984 and Pesaran and Smith, 1990) where the structural model is completed by marginal models for the variables whose exogeneity is questioned. For this purpose, Hausman-type (1978) specification tests appear as the more appealing procedures, at least from a practitioner's point of view. Simulation evidence reported in Kiviet (1986) does nevertheless point out the rather poor behaviour of orthogonality tests in samples of typical size used in macroeconometric applications. In the case of nonstationary explanatory variables, Krämer (1985) has investigated the asymptotic properties of the Hausman test in the presence of (deterministically) trended explanatory variables and showed that after proper normalisation, the asymptotic null distribution of the Hausman test remains standard. He did however not consider the behaviour of the Hausman test when the variables are generated by integrated processes. This particular point will be considered in this chapter whose structure is as follows.

We first present the way traditional exogeneity tests (in a stationary framework) are performed and the implicit model representation of the SEM on which they are based. We then consider the behaviour of orthogonality tests in the case of cointegrated variables. This is investigated by means of a small scale *illustrative* Monte Carlo experiment. The limiting distribution of the Hausman test and the conditions under which the usual χ^2 limiting distribution holds are investigated. Finally we also discuss a simple extension of orthogonality tests along the lines of Boswijk's (1991b) test in order to test for weak exogeneity when both long and short run dynamic response parameters are of interest.

4.2 Exogeneity and the Incomplete Simultaneous Equation Model

The first question that comes to mind when one is confronted with weak exogeneity tests,[1] is "in what model should we analyse exogeneity, and in what model representation should we work for testing this important hypothesis?".

In the stationary framework, this problem has recently been considered by Bauwens and van Dijk (1990) in their study of Bayesian limited information analysis. In general, weak exogeneity as defined by Engle, Hendry and Richard (1983) has often been questioned for requiring knowledge of the extent to which single equation analysis is valid or whether higher dimensional models should be used. Since the work of Richard (1980, 1984), Engle (1984), Holly (1985), Engle et al. (1983) and others, the concept of exogeneity has been investigated in a framework labelled *incomplete* simultaneous equation models, i.e. models where the number of potential endogenous variables is greater than the number of structural equations. To quote Richard (1979):

> *....complete simultaneous equation models should be the exception and no econometricians should ever feel obliged to exogenize variables and/or add equations to a model just for the purpose of completing it.*

We will however see that different representations of the incomplete SEM are available, and that in the case of cointegrated systems, a particular form emerges as potentially more useful for empirical analysis.

Let us first discuss the problem in the stationary framework using standard matrix notation for the presentation of SEMs. We consider the case of

[1] We must note that some authors, for example Geweke (1984), have argued that weak exogeneity *per se* is not a testable concept as it does not generate refutable hypotheses and is moreover based on the choice of parameters of interest, and thus is a subjective notion. Horvath (1991) on the other hand argues in favour of linking the various exogeneity concepts to more easily testable hypothesis not needing the construction of marginal models. We shall not follow these approaches and instead shall recognize the possibility of testing for exogeneity via the parametric restrictions that it implies in the context of particular model representations (see Holly, 1985; Bauwens and van Dijk, 1990).

a single structural equation (possibly dynamic) taken from a complete SEM where we have a *g-vector* of potential endogenous variables which we denote by Y_1.

$$y = Y_1\alpha + X_1\beta + u \tag{4.1}$$

where y and Y_1 are respectively a $T \times 1$ vectors and a $T \times g$ matrix of observations on the endogenous variables and the "potential" endogenous variables in Y_1. X_1 is a $T \times k_1$ matrix of observations on k_1 predetermined variables present in the equation for y. We denote by X_2 a $T \times k_2$ matrix of excluded predetermined variables which along with X_1 form the $(T \times (k_1 + k_2)$ matrix $X = (X_1, X_2)$. We assume that the variables in X_2 are valid instruments, i.e. $E(u_t X_{2t-i}) = 0$, $i = 0, 1, 2, \ldots$. The parameters of interest are those of the structural equation (4.1).

The reduced form for Y_1 is

$$Y_1 = X\Pi + V \tag{4.2}$$

where V is a $T \times g$ vector of unobserved reduced form disturbances and Π is a $(k_1 + k_2) \times 1$ vector of reduced form parameters. We also assume a joint normal density for $(u, V)'$ with p.d.s. covariance matrix Σ:

$$\begin{pmatrix} u_t \\ V_t \end{pmatrix} \sim N\left[\begin{pmatrix} 0 \\ 0 \end{pmatrix}, \begin{pmatrix} \sigma_{11} & \sigma_{12} \\ \sigma_{21} & \Sigma_{22} \end{pmatrix}\right] \tag{4.3}$$

where $\sigma_{12} = \sigma_{21}'$. Under these conventions, the log likelihood function of the model (4.1, 4.2) is proportional to

$$L \propto -\frac{T}{2}\log|\Sigma| - \frac{1}{2}\operatorname{tr}\Sigma^{-1}S \tag{4.4}$$

where

$$S = \begin{pmatrix} u'u & u'V \\ V'u & V'V \end{pmatrix} \tag{4.5}$$

In order to point out the potential model representations under which one can analyse exogeneity, we consider the two possible ways of writing down the factorisation of this likelihood function. As usual we express it in terms of the disturbances $(u\ V')$ and consider the fact that $D(u, V) = D(u|V)\,D(V) =$

$D(V|u)\,D(u)$. Note that both factorisations are obtained in a straightforward manner by considering the inversion of a partitioned matrix with positive definite diagonal blocks. Using the rule of inversion of a partitioned matrix we may write Σ^{-1} as (see Gourieroux and Monfort, 1989):

$$\Sigma^{-1} = \begin{pmatrix} \sigma_{11}^{-1} + \sigma_{11}^{-1}\sigma_{12}\eta^{-1}\sigma_{21}\sigma_{11}^{-1} & -\sigma_{11}^{-1}\sigma_{12}\eta^{-1} \\ -\eta^{-1}\sigma_{21}\sigma_{11}^{-1} & \eta^{-1} \end{pmatrix} \tag{4.6}$$

where $\eta = \Sigma_{22} - \sigma_{21}\sigma_{11}^{-1}\sigma_{12}$ and det $(\Sigma) = \mid \Sigma \mid = \mid \sigma_{11} \mid\mid \eta \mid$.In this case the log-likelihood function is proportional to

$$\begin{aligned} L \;\propto\; & -\frac{T}{2}\log\sigma_{11} - \frac{u'u}{2\sigma_{11}} - \frac{T}{2}\log\mid\eta\mid \\ & -\frac{1}{2}\mathrm{tr}(\eta^{-1}(V - \frac{u'\sigma_{21}}{\sigma_{11}})'(V - \frac{u\sigma_{12}}{\sigma_{11}})) \end{aligned} \tag{4.7}$$

This factorisation is favoured and extensively used by Richard (1980, 1984), Steel (1987a, 1987b) and Steel and Richard (1991) . The first two terms correspond to the marginal density $D(y)$ while the remaining two terms are those corresponding to $D(Y_1|y)$. Notice that the nuisance parameter σ_{12} only appears in the latter part of the factorisation which is an interesting property for the derivation of LR tests of weak exogeneity (see Richard, 1979, 1984).

However, for practical purposes we may also consider the alternative factorisation which can easily be obtained with the alternative form of Σ^{-1}:

$$\Sigma^{-1} = \begin{pmatrix} \omega^{-1} & -\sigma_{12}\Sigma_{22}^{-1}\omega^{-1} \\ -\Sigma_{22}^{-1}\omega^{-1}\sigma_{21} & \Sigma_{22}^{-1} + \Sigma_{22}^{-1}\sigma_{21}\omega^{-1}\sigma_{12}\Sigma_{22}^{-1} \end{pmatrix} \tag{4.8}$$

and $\det(\Sigma) = \mid \Sigma \mid = \mid \omega \mid\mid \Sigma_{22} \mid$ with $\omega = \sigma_{11} - \sigma_{12}\Sigma_{22}^{-1}\sigma_{21}$. With this expression for Σ^{-1}, the log-likelihood function becomes proportional to

$$\begin{aligned} L \propto \frac{T}{2}\log\omega - \frac{T}{2}\omega^{-1}(u - V\Sigma_{22}^{-1}\sigma_{21})'(u - V\Sigma_{22}^{-1}\sigma_{21}) \\ -\frac{T}{2}\log\mid\Sigma_{22}\mid - \frac{1}{2}\mathrm{tr}\Sigma_{22}^{-1}V'V \end{aligned} \tag{4.9}$$

Here, the first two terms corresponds to the *conditional* distribution of $y|Y_1$ while the last two to the *marginal* of Y_1 and thus it implies the factorisation $D(y, Y_1) = D(y|Y_1)D(Y_1)$. Since we are working in a Gaussian framework, the independence between Y_1 and u is equivalent to $\text{cov}(Y_1, u) = \text{cov}(V, u) = \sigma_{12} = 0$. Testing for weak exogeneity of Y_1 for the parameters of interest (α, β) corresponds to the parametric restriction $H_0 : \sigma_{12} = 0$ against $H_a : \sigma_{12} \neq 0$. Both factorisations are in one-to-one correspondence with Σ. This correspondence between Σ, $(\sigma_{11}, \lambda, \eta)$, $(\Sigma_{22}, \mu, \omega)$ with $\mu = \sigma_{12}\Sigma_{22}^{-1}$, $\lambda = \sigma_{21}\sigma_{11}^{-1}$ is given by

$$\Sigma = \begin{pmatrix} \sigma_{11} & \sigma_{12} \\ \sigma_{21} & \Sigma_{22} \end{pmatrix} = \begin{pmatrix} \sigma_{11} & \lambda\sigma_{11} \\ \lambda\sigma_{11} & \eta - \lambda^2\sigma_{11} \end{pmatrix} = \begin{pmatrix} \omega + \mu^2\Sigma_{22} & \mu\Sigma_{22} \\ \mu\Sigma_{22} & \Sigma_{22} \end{pmatrix} \tag{4.10}$$

One can thus alternatively test for $\lambda = 0$ or $\mu = 0$. Following an artificially expanded regression framework, $\hat{V}$ is added to (4.1) under the second model representation (see Holly, 1983, 1985, Engle, 1984) while in the other case, $\hat{u}$ is added to (4.2) (see Steel and Richard, 1991, Capuccio and Orsi, 1990).

These two alternative factorisations imply two different representations of the incomplete simultaneous equation model within which the hypothesis of weak exogeneity of Y_1 for the parameters of interest (α, β) can be tested. These are summarised in Table 4.1, adapted from Bauwens and van Dijk (1990).

Model Representation	I	II
Structural Form Reduced Form	u remains $\hat{u}$ added to (4.2)	$\hat{V}$ added to (4.1) V remains

Table 4.1: Representation of incomplete SEMs

Model representation I is favoured in the work of Richard (1980, 1984), Steel (1987, 1987b). Model representation II is favoured on the other hand by

Holly (1983, 1985) and Engle (1984) who derive LM tests for weak exogeneity. Note the obvious disadvantage that, although emphasis is on conditional sub-models, marginal models have to be specified and fitted if weak exogeneity is to be tested, leaving open the danger of misspecifying the marginal generating process. In other words, we need to specify a marginal model in order to know ex-post whether it is needed or not for efficient inference.[2] Different approaches are naturally feasible leading to more or less restrictive specifications of the marginal models.

We can naturally derive a complete set of structural equations for each variable in Y_1 so that the analysis would be placed in a *full information* set-up. This is however often inefficient and computationally demanding since economic theory usually provides little guidance for this type of complete structural specification. Therefore, different degrees of flexibility can be achieved by considering:

- the least flexible, the *limited information* approach, in which all variables retained in (4.2) are used as instruments, i.e. Π in (4.2) is left unrestricted;
- a framework where some of the elements of X may be dropped, but this for *all* elements in Y_1;
- a more flexible approach where the instruments are selected according to their own merits, i.e. linear restrictions are incorporated on the coefficients of Π see Lubrano et al. (1986), Steel (1987b) and Zellner et al. (1988).

The differences among these three levels of flexibility are related to the way overidentifying restrictions are handled. In the least flexible, the *limited*

[2]An indirect way of addressing the problem of weak exogeneity is through the implication of super exogeneity in which case a within sample stability analysis can prove to be useful. Effectively, as it has been argued in Hendry (1987), if a conditional model has constant within sample parameters although it is known that the processes generating the regressors have been subject to major regime changes, then we may intuitively accept the hypothesis of super exogeneity of the regressors, and thus weak exogeneity. In the case of cointegrated systems, as has appeared clear in the previous chapter, the cross-equation restriction implied by the (potential) presence of the cointegrating vector in the remaining equation is such that these indirect ways of checking weak exogeneity should be used only if it is known that the cointegrating vector is absent from the marginal model.

information approach, they are all deleted and the structural equation is completed by purely *unrestricted* reduced forms while in the GIV framework, exclusion restrictions (such as Granger non-causality claims) can be imposed on the reduced form matrix Π.[3]

4.3 The Behaviour of Orthogonality Tests in the Presence of (Co)integrated Variables

Let us now consider the behaviour of these orthogonality tests in cointegrated systems.[4] Although we know that the orthogonality hypothesis is not *sufficient* for weak exogeneity it remains useful for any assessment of weak exogeneity if the cointegrating vector is known to be absent from the marginal models. As pointed out in the introduction to this chapter, an important problem which is likely to arise with (co)integrated processes in practical applications of orthogonality tests is a nonstandard limiting distribution.

Consider a simple example which is of practical relevance in empirical applications of orthogonality tests. Reduced form models for the potential endogenous variable, say y_{1t}, are often specified by including variables in levels, although they are $I(1)$ and possibly not cointegrated with y_{1t}. Examples include Lubrano et al. (1986), Steel (1987a, 1987b) and Urbain (1988). In this case, the estimated residual used in the augmented model might well be $I(1)$ so that the behaviour of the orthogonality tests is potentially nonstandard.[5]

An interesting question is that of the behaviour, in terms of small sample power and size, of variable addition exogeneity tests when a traditional χ^2 distribution is used, and this particularly for various sets of the instruments, covering balanced and unbalanced cases. In order to point out some important features of orthogonality tests in this framework, we shall first make use of a small scale Monte Carlo experiment. The asymptotic distribution

[3] At our knowledge, no studies have yet compared these alternative approaches in finite samples.

[4] This section is based on a joint work with Peter Boswijk, see Boswijk and Urbain (1991).

[5] This problem is likely to hold for many other variable addition tests, but this generalisation goes far outside the scope of this chapter.

of the Hausman test in the context of the DGP used in the simulations is also considered.

4.3.1 Testing for weak exogeneity within a limited information framework

Since Wu's (1973) study, an important number of orthogonality tests have been proposed in the literature. As shown for example in Holly (1983, 1985), most of these can be reformulated in a unified expanded regression framework. See also Nakamura and Nakamura (1981) who show the asymptotic equivalence of some of them. They basically only differ by the use of the estimator of the disturbance variance. Asymptotically all of these variance estimators are consistent and equivalent under the null, but as usual they might produce quite different numerical values in small samples such as those typically used with quarterly or annual time series data.

Among the numerous alternative forms of orthogonality tests that have been proposed we restrict our attention to those which were shown to perform the "less poorly" in the detailed Monte Carlo study of Kiviet (1986) who considered the behaviour of a number of misspecification tests in the case of a single equation taken from a complete dynamic SEM. We therefore focus on Wu's (1973) T2 test used in F-form, the χ^2 version given in Hausman (1978) and the corrected version of Wu's T4 test statistic proposed by Nakamura and Nakamura (1981). They are respectively denoted by **W**, **H** and **N**.

Consider for simplicity the case of a single structural equation (possibly dynamic) taken from a complete SEM.

$$y = X_1\beta_1 + x_2\beta_2 + u \tag{4.11}$$

where y and x_2 are $T \times 1$ vectors of observation on the endogenous variables y and the "potential" endogenous variables x_2. X_1 is a $T \times k_1$ matrix of observations on k_1 predetermined variables present in the equation for y. We denote by Z_2 a $T \times l_2$ matrix of excluded predetermined variables which along with X_1 form the $T \times (k_1 + l_2) = T \times l$ matrix of instruments $Z = (X_1, Z_2) = (Z_1, Z_2)$ with $l = l_1 + l_2 = k_1 + l_2$. We also maintain the variables in Z_2 as valid instruments, i.e. $E(u_t Z_{2t-i}) = 0$, $i = 0, 1, 2, \ldots.$

The reduced form for x_2 is

$$x_2 = Z\Pi + v \tag{4.12}$$

where v is a $(T \times 1)$ vector of reduced form disturbances and Π is a $(l \times 1)$ vector of reduced form parameters. We assume a joint normal density for $(u, v)'$:

$$\begin{pmatrix} u_t \\ v_t \end{pmatrix} \sim N\left[\begin{pmatrix} 0 \\ 0 \end{pmatrix}\begin{pmatrix} \sigma_{11} & \sigma_{12} \\ \sigma_{21} & \sigma_{22} \end{pmatrix}\right] \tag{4.13}$$

The null hypothesis of orthogonality, which in this framework is a sufficient condition for the weak exogeneity of x_2 w.r.t. the parameters of (4.11), is given by $H_0 : \sigma_{12} = 0$. A simple way to proceed is to estimate (4.12) in order to obtain the OLS estimates $\hat{v}$. If we add $\hat{v}$ to the single equation structural model (4.11) model, we get the artificially *expanded* model

$$y = x_1\beta_1 + X_2\beta_2 + \hat{v}\delta + u \tag{4.14}$$

Since $\delta = \sigma_{12}\sigma_{22}^{-1}$, the null hypothesis of orthogonality becomes $H_0 : \hat{\delta} = 0$ against $H_a : \hat{\delta} \neq 0$. Testing for weak exogeneity of x_2 for (β_1, β_2) is then simply testing for omitted variables. The tests statistics we consider are given by:

$$\mathbf{H} = T\,\frac{RRSS - URSS}{URSS} \tag{4.15}$$

$$\mathbf{W} = \frac{(T-k-k_2)}{k_2}\,\frac{RRSS - URSS}{URSS} = \frac{(T-k-k_2)}{Tk_2}\,\mathbf{H} \tag{4.16}$$

$$\mathbf{N} = \frac{(T-k)}{k_2}\,\frac{RRSS - URSS}{RRSS} = \frac{(T-k)}{k_2}\,\frac{\mathbf{H}}{\mathbf{H}+T} \tag{4.17}$$

where T is the number of observations, k the number of explanatory variables variables under the null hypothesis, k_2 is the number of variables whose exogeneity is questioned (in our case $k_2 = 1$). $RSS(URSS)$ is the OLS residual sum of squares of the restricted model (the unrestricted expanded model). $\mathbf{W}$, equivalent to Wu's T_2 test (see Kiviet, 1986), is distributed as an F $(k_2, T-k-k_2)$; **H** is the χ^2 version of T_2 due to Hausman (1978), distributed as a $\chi^2(k_2)$ and **N**, distributed as an $F(k_2, T-k)$, is the modification of Wu's T_4 test proposed by Nakamura and Nakamura (1981).[6]

4.3.2 Small sample behaviour: some simulation evidence

Our experiment is based on a very simple data generating process motivated by empirical studies in the "Hendry" framework. It consists of a single equation error correction model (in structural form) completed by simple generating models for the explanatory variables, the potential endogenous variable and an external instrument not included in the structural equation.

The DGP

We consider the following structural error correction model taken as the true data generating model for y_t

$$\Delta y_t = \beta_1 \Delta x_{1t} + \beta_2 \Delta x_{2t} + \gamma(y_{t-1} - \theta_1 x_{1,t-1} - \theta_2 x_{2,t-1}) + u_{1t} \tag{4.18}$$

where (y_t, x_{1t}, x_{2t}) are $I(1)$ cointegrated variables, denoted by $CI(1,1)$, with a unique cointegrating vector $(1, -\theta_1, -\theta_2)'$ and the disturbances u_t are i.i.d. $N(0, \sigma_{11})$. We assume that this model is dynamically well-specified, but that x_{2t} may not be weakly exogenous for the parameters of interest which include both the long run and short run parameters : $(\beta_1, \beta_2, \gamma, \theta_1, \theta_2)$. Single equation methods such as OLS may therefore yield inconsistent or inefficient estimates.

In order to simplify the analysis, we first assume that y_t is the only correcting variable, that x_{1t} is weakly exogenous for the parameters of interest and that the joint distribution of (x_{1t}, x_{2t}) conditional upon the past information set does not involve the parameters (θ_1, θ_2).

$$\Delta x_{1t} = u_{2t} \tag{4.19}$$

$$\Delta x_{2t} = u_{3t} \tag{4.20}$$

[6]In the notation of Kiviet (1986), they are denoted by ORF2, OR2 and ORF4 respectively.

Although this last assumption is stated here for simplification purposes it is necessary since the presence of the cointegrating vector in the marginal model with non-zero error correction coefficient is sufficient to reject the null hypothesis of weak exogeneity. As a matter of illustration we also performed the same set of experiments when we allow the cointegrating vector to enter the generating model for x_{2t} so that (4.20) is replaced by

$$\Delta x_{2t} = \mu(y_{t-1} - \theta_1 x_{1,t-1} - \theta_2 x_{2,t-1}) + u_{3t} \tag{4.21}$$

with $\mu = 0.10$. In this case however, orthogonality is not sufficient for weak exogeneity (see Proposition 4 in the previous Chapter). It is, however, interesting to see to what extend this small modification of the DGP alters the experimental results since this case is often of empirical relevance.

Finally we have to assume a simple generating model for an external instrument, denoted by z_t, correlated with x_{2t} but not with y_t. Its validity is not in doubt so that we impose $E(u_{1t}z_{t-j}) = 0$ for $j = 0, 1, \ldots$

$$z_t = z_{t-1} + u_{4t} \tag{4.22}$$

and the vector of disturbances will be generated according to a multivariate normal distribution with zero mean vector and covariance matrix given by

$$\Sigma = \begin{pmatrix} \sigma_{11} & 0 & \sigma_{13} & 0 \\ 0 & \sigma_{22} & 0 & 0 \\ \sigma_{13} & 0 & \sigma_{33} & \sigma_{34} \\ 0 & 0 & \sigma_{34} & \sigma_{44} \end{pmatrix} \tag{4.23}$$

It must be pointed out that our framework does consider an external instrument not cointegrated with x_{2t}, or with y_t. We do also not allow x_{1t} and x_{2t} to be cointegrated without y_t. A similar DGP has recently been investigated by Dolado et al. (1989) who pointed out that when we allow for example x_{1t} and x_{2t} to be cointegrated alone (i.e. without y_t), we encounter near perfect asymptotic collinearity which results in the loss of mixed normality of the least squares estimates of the cointegrating vector components in (4.18).[7]

Hypotheses concerning the weak exogeneity of x_{2t} are easily formulated since, in our DGP, x_{2t} is weakly exogenous for the parameters of interest

[7]This is obvious since in this case the rank of $(X'X)$ will tend to 1

if and only if $\sigma_{13} = 0$. Note that $\sigma_{34} = 0$ would make z_t a "spurious" instrument. The following parameter values have been taken for our design

- T=50;
- $\sigma_{22} = \sigma_{33} = \sigma_{44}$; and we use two values for the ratio $\sigma_{11}/\sigma_{33} = \{1/3, 3/1\}$;
- $\text{cor}(\Delta x_{2t}, u_{1t}) = \rho = \{0.0, 0.25, 0.50\}$;
- $(\beta_1, \beta_2) = (0.50, 0.10)$; $(\theta_1, \theta_2) = (1.0, 0.50)$
- the error correction coefficient is allowed to take the following values: $\gamma = \{-0.10, -0.50\}$;
- $\text{cor}(\Delta x_{2t}, z_t) = 0.60$ for all experiments.

For each experiment, we performed 500 replications on a MacIntosh SE/30 using the RATS 3.11. The particular choice of these parameters values was suggested by long and short run coefficient estimates of aggregate import demand studies in Urbain (1990). Given the small sample size usually available in applied econometrics (especially with time series data), we limit the presentation[8] to T=50. In each replication, we estimate by ordinary least squares a single equation error correction model

$$\Delta y_t = b_1 \Delta x_{1t} + b_2 \Delta x_{2t} + b_3 y_{t-1} + b_4 x_{1,t-1} + b_5 x_{2,t-1} + u_{1t} \qquad (4.24)$$

and compute Hausman, Wu and Nakamura tests for weak exogeneity of x_{2t} for $(\beta_1, \beta_2, \gamma, \theta_1, \theta_2)$ using some instruments. In practice we may consider two major cases

1. cases where a reduced form model for the level of x_{2t} is fitted
2. cases where a reduced form model for the first difference Δx_{2t} is fitted.

For each instrument choice, the case where a model for the level is fitted is denoted by the subscript A, while the model with differences is denoted by a subscript B. We choose several different sets of instruments which are

[8] In practice we generated samples of 70 observations and dropped the first 20 observations in order to avoid startup dependencies.

presented in Table 4.2. The motivation for estimating a marginal model for the level as well as for the first difference stems from the way empirical applications are performed. For example, the explicit modelling of the marginal model is often based on a auxiliary model for the first differences (see Steel and Richard, 1991; Lubrano et al., 1986). The instrument sets reported in Table 4.2 are broadly consistent with those used in some of the empirical examples mentioned above. In particular it can be noted that the sets 1, 2 and 4 imply Granger non-causality claims from y_t to x_{2t}.

Case	Variables
set 1_A	$(constant, x_{1t}, x_{1t-1}, x_{2t-1})$
set 1_B	$(constant, \Delta x_{1t}, \Delta x_{1t-1}, \Delta x_{2t-1})$
set 2_A	$(constant, x_{1t}, x_{1t-1}, z_t, z_{t-1})$
set 2_B	$(constant, \Delta x_{1t}, \Delta x_{1t-1}, \Delta z_t, \Delta z_{t-1})$
set 3_A	$(constant, \Delta x_{1t}, x_{1t-1}, \Delta z_t, z_{t-1}, y_{t-1})$
set 3_B	$(constant, \Delta x_{1t}, x_{1t-1}, \Delta z_t, z_{t-1}, y_{t-1})$
set 4_A	$(constant, x_{1t}, x_{1t-1}, x_{2t-1}, z_t, z_{t-1})$
set 4_B	$(constant, \Delta x_{1t}, \Delta x_{1t-1}, \Delta x_{2t-1}, \Delta z_t, \Delta z_{t-1})$

Table 4.2: Instruments used in the simulations

Simulation results

The following three Tables report the rejection rates obtained during the simulations. For the estimated size (type 1 probability error) we shall say that the actual size differs significantly from the nominal 0.05 level only if it is outside the interval $0.05 \pm 2(\sqrt{Rr(1-r)})/R$ where R is the total number of replications (500) and r is the fraction of rejections. In our case this means that the bounds are respectively 0.033 and 0.073.

The empirical size are reported in Table 4.3. The numbers in brackets are the rejection frequencies obtained when the auxiliary model is the model for the first difference of x_{2t}, i.e. it corresponds to the various cases denoted with a subscript "B" in Table 4.2.

In accordance with the simulation outcome reported in Kiviet (1986) the addition of an external exogenous instrument (e.g. z_t) does not significantly improve the empirical size of the various orthogonality tests as can be seen in Table 4.3 where the empirical size obtained from the rejection rates with instruments set 1_A and 1_B are not significantly worse than those obtained with an external instrument (cases 2,3,4). Note that for almost all parameter configurations, the second instruments set rejects the correct null too often: up to 80% of the replications for the ratio $\sigma_{11}/\sigma_{33} = 3/1$, i.e. when the potential endogenous variable x_{2t} has a lower variability than y_t. Similar important overrejection rates are observed for all instruments sets when $\sigma_{11}/\sigma_{33} = 3/1$ and when the cointegrating vector does not appear in the generating model of x_{2t}, i.e. when $\mu = 0$. In this case, fitting an auxiliary model for the first differences leads to a rejection rate of the correct null more often than when the level is modelled as shown for example in cases 3_A and 3_B which share the same instruments. Note also that when the x_{2t} is error correcting ($\mu \neq 0$) the size distortion is usually smaller when the first difference is modelled while the converse seems to hold, at least for instruments sets 2,3 and 4. Overall however, important size distortions are observed in almost all cases. Either the significance level is too low or the empirical size is dramatically high. Concerning the individual performances, **H** usually displays the highest size distortion while **N** and **W** have very similar behaviour.

The empirical powers are reported in Table 4.4 and 4.5. In Table 4.4, the results are reported when the correlation between x_{2t} and u_{1t} is fixed in the true DGP at 0.25, while in Table 4.5 it equals 0.50. First of all, and irrespective of the presence or absence of the cointegrating vector in the generating equation of x_{2t}, there is no significant increase in the empirical power when the correlation increases from 0.25 to 0.50.

Contrary to what was observed in the previous size analysis, an instruments set without an external exogenous instrument not already included in the structural model, i.e. sets 1_A and 1_B, performs the worst, except when x_{2t} is error correcting, when $\sigma_{11}/\sigma_{33} = 1/3$ and when the error correction

Inst. set	γ	Tests	$\sigma_{11}/\sigma_{33} = 1/3$		$\sigma_{11}/\sigma_{33} = 3/1$	
			$\mu = 0$	$\mu = 0.10$	$\mu = 0$	$\mu = 0.10$
$1_A(1_B)$	0.50	W	0.00 (0.11)	0.00 (0.13)	0.08 (0.10)	0.01 (0.02)
		H	0.00 (0.14)	0.00 (0.17)	0.11 (0.12)	0.01 (0.04)
		N	0.00 (0.09)	0.00 (0.13)	0.06 (0.08)	0.01 (0.02)
	0.10	W	0.00 (0.09)	0.00 (0.12)	0.24 (0.12)	0.45 (0.32)
		H	0.00 (0.14)	0.01 (0.18)	0.26 (0.14)	0.50 (0.32)
		N	0.00 (0.07)	0.00 (0.11)	0.20 (0.12)	0.40 (0.31)
$2_A(2_B)$	0.50	W	0.18 (0.10)	0.47 (0.06)	0.74 (0.83)	0.24 (0.11)
		H	0.22 (0.14)	0.53 (0.09)	0.83 (0.85)	0.27 (0.13)
		N	0.16 (0.06)	0.44 (0.06)	0.68 (0.83)	0.22 (0.11)
	0.10	W	0.09 (0.28)	0.50 (0.23)	0.70 (0.84)	0.14 (0.11)
		H	0.13 (0.35)	0.53 (0.24)	0.74 (0.88)	0.16 (0.15)
		N	0.06 (0.25)	0.47 (0.23)	0.66 (0.81)	0.10 (0.08)
$3_A(3_B)$	0.50	W	0.08 (0.42)	0.15 (0.06)	0.10 (0.52)	0.00 (0.00)
		H	0.09 (0.48)	0.20 (0.09)	0.15 (0.58)	0.00 (0.00)
		N	0.08 (0.42)	0.13 (0.06)	0.10 (0.48)	0.00 (0.00)
	0.10	W	0.19 (0.44)	0.50 (0.23)	0.13 (0.57)	0.03 (0.01)
		H	0.25 (0.47)	0.46 (0.24)	0.18 (0.57)	0.03 (0.01)
		N	0.19 (0.43)	0.47 (0.23)	0.13 (0.51)	0.03 (0.01)
$4_A(4_B)$	0.50	W	0.38 (0.46)	0.02 (0.02)	0.57 (0.80)	0.00 (0.00)
		H	0.38 (0.47)	0.04 (0.03)	0.62 (0.85)	0.00 (0.00)
		N	0.34 (0.46)	0.02 (0.02)	0.54 (0.78)	0.00 (0.00)
	0.10	W	0.32 (0.41)	0.09 (0.08)	0.54 (0.73)	0.00 (0.00)
		H	0.37 (0.48)	0.12 (0.10)	0.58 (0.78)	0.00 (0.00)
		N	0.31 (0.39)	0.08 (0.06)	0.52 (0.73)	0.00 (0.00)

Table 4.3: Empirical size of orthogonality tests

Inst. set	γ	Tests	$\sigma_{11}/\sigma_{33} = 1/3$		$\sigma_{11}/\sigma_{33} = 3/1$	
			$\mu = 0$	$\mu = 0.10$	$\mu = 0$	$\mu = 0.10$
$1_A(1_B)$	0.50	W	0.01 (0.1)	0.00 (0.13)	0.09 (0.08)	0.05 (0.00)
		H	0.03 (0.14)	0.01 (0.15)	0.11 (0.11)	0.06 (0.03)
		N	0.01 (0.09)	0.00 (0.11)	0.08 (0.07)	0.04 (0.00
	0.10	W	0.00 (0.15)	0.00 (0.12)	0.23 (0.15)	0.45 (0.31)
		H	0.00 (0.20)	0.00 (0.16)	0.30 (0.19)	0.47 (0.36)
		N	0.00 (0.14)	0.00 (0.07)	0.18 (0.13)	0.44 (0.27)
$2_A(2_B)$	0.50	W	0.21 (0.06)	0.36 (0.05)	0.67(0.81)	0.21 (0.12)
		H	0.24 (0.09)	0.46 (0.06)	0.75 (0.83)	0.25 (0.19)
		N	0.19 (0.06)	0.33 (0.04)	0.66 (0.81)	0.20 (0.11)
	0.10	W	0.10 (0.28)	0.45 (0.24)	0.67 (0.87)	0.12 (0.08)
		H	0.16 (0.35)	0.52 (0.31)	0.75 (0.89)	0.17 (0.14)
		N	0.08 (0.26)	0.44 (0.22)	0.65 (0.87)	0.10 (0.07)
$3_A(3_B)$	0.50	W	0.11 (0.46)	0.18 (0.13)	0.09 (0.52)	0.00 (0.00)
		H	0.12 (0.51)	0.20 (0.18)	0.11 (0.57)	0.00 (0.00)
		N	0.09 (0.43)	0.17 (0.13)	0.08 (0.50)	0.00 (0.00)
	0.10	W	0.18 (0.50)	0.47 (0.18)	0.09 (0.60)	0.07 (0.03)
		H	0.22 (0.53)	0.48 (0.25)	0.11 (0.67)	0.11 (0.03)
		N	0.17 (0.50)	0.46 (0.18)	0.09 (0.57)	0.06 (0.03)
$4_A(4_B)$	0.50	W	0.48 (0.48)	0.06 (0.03)	0.49 (0.65)	0.00 (0.01)
		H	0.51 (0.53)	0.08 (0.04)	0.61 (0.69)	0.00 (0.01)
		N	0.44 (0.45)	0.06 (0.02)	0.45 (0.58)	0.00 (0.00)
	0.10	W	0.51 (0.51)	0.16 (0.07)	0.51 (0.57)	0.00 (0.00)
		H	0.54 (0.56)	0.20 (0.15)	0.55 (0.64)	0.01 (0.01)
		N	0.48 (0.51)	0.14 (0.06)	0.48 (0.54)	0.01 (0.00)

Table 4.4: Empirical power of orthogonality tests , $\rho = 0.25$

Inst. set	γ	Tests	$\sigma_{11}/\sigma_{33} = 1/3$		$\sigma_{11}/\sigma_{33} = 3/1$	
			$\mu = 0$	$\mu = 0.10$	$\mu = 0$	$\mu = 0.10$
$1_A(1_B)$	0.50	W	0.01 (0.10)	0.02 (0.12)	0.06 (0.07)	0.04 (0.03)
		H	0.02 (0.15)	0.03 (0.16)	0.10 (0.13)	0.05 (0.04)
		N	0.01 (0.06)	0.01 (0.10)	0.06 (0.07)	0.03 (0.03)
	0.10	W	0.00 (0.07)	0.01 (0.17)	0.15 (0.18)	0.43 (0.28)
		H	0.01 (0.12)	0.01 (0.21)	0.20 (0.19)	0.45 (0.31)
		N	0.01 (0.15)	0.01 (0.15)	0.14 (0.18)	0.41 (0.27)
$2_A(2_B)$	0.50	W	0.19 (0.08)	0.45 (0.08)	0.50 (0.80)	0.22 (0.07)
		H	0.23 (0.17)	0.48 (0.09)	0.55 (0.81)	0.24 (0.07)
		N	0.19 (0.08)	0.43 (0.07)	0.48 (0.78)	0.21 (0.07)
	0.10	W	0.17 (0.29)	0.48 (0.19)	0.56 (0.86)	0.06 (0.08)
		H	0.20 (0.33)	0.50 (0.20)	0.66 (0.87)	0.10 (0.12)
		N	0.15 (0.28)	0.47 (0.18)	0.54 (0.86)	0.06 (0.06)
$3_A(3_B)$	0.50	W	0.16 (0.50)	0.13 (0.08)	0.06 (0.47)	0.00 (0.00)
		H	0.17 (0.52)	0.22 (0.11)	0.09 (0.49)	0.00 (0.00)
		N	0.11 (0.48)	0.14 (0.06)	0.06 (0.43)	0.00 (0.00)
	0.10	W	0.14 (0.41)	0.47 (0.20)	0.09 (0.42)	0.06 (0.03)
		H	0.17 (0.47)	0.51 (0.28)	0.12 (0.46)	0.07 (0.04)
		N	0.14 (0.40)	0.46 (0.19)	0.09 (0.42)	0.06 (0.03)
$4_A(4_B)$	0.50	W	0.37 (0.45)	0.02 (0.01)	0.38 (0.46)	0.00 (0.01)
		H	0.38 (0.48)	0.02 (0.04)	0.41 (0.52)	0.00 (0.01)
		N	0.34 (0.40)	0.02 (0.00)	0.45 (0.38)	0.00 (0.00)
	0.10	W	0.37 (0.38)	0.15 (0.04)	0.34 (0.41)	0.00 (0.00)
		H	0.39 (0.44)	0.17 (0.19)	0.41 (0.46)	0.00 (0.01)
		N	0.35 (0.38)	0.15 (0.04)	0.32 (0.36)	0.00 (0.00)

Table 4.5: Empirical power of orthogonality tests , $\rho = 0.50$

coefficient is low. The remaining results are, in any case, not very favourable. As far as the presence of the cointegrating vector is concerned, its presence in the generating model of x_{2t} has dramatic effects on the estimated power: almost zero for the sets 3 and 4 when the ratio σ_{11}/σ_{33}=3/1 and for the sets 1 and 4 when it equals 1/3. The latter parameter (the ratio σ_{11}/σ_{33}) has an important effect. In particular, when it equals 1/3 and when x_{2t} is not error correcting, the estimated power is almost always lower than for the reverse ratio. A possible explanation of this result can eventually be found if we consider how these orthogonality tests work. Basically they are testing the significance of the added fitted residual (obtained from fitting an auxiliary model for x_{2t}) in (4.18). This parameter is given by $\delta = \sigma_{13}/\sigma_{33}$. The higher the variability of x_{2t} w.r.t. to that of y_t, for a given correlation between x_{2t} and u_{1t}, the lower will δ be, probably inducing a lower rejection rate of the incorrect null hypothesis $\sigma_{13} = 0$. When the cointegrating vector appears in the generating model for x_{2t}, the latter effects seem to disappear.

However, we should not draw too rapid conclusions from these results given (*i*) the limited range of cases considered, (*ii*) the non-standard limiting distribution that is likely to apply in some cases. What they at least point out, is that orthogonality tests have poor behaviour for parameter settings and models such as those we have considered in this small scale illustrative experiment. Drawing more general conclusions would require a more extensive simulation exercise which is outside the scope of this work. According to our limited simulation evidence, it appears anyway that the behaviour of these tests is extremely bad, even worse than those reported in Kiviet (1986). In most cases, the empirical powers are effectively not much greater than the (distorted) empirical size obtained under the correct null of orthogonality.

4.3.3 Asymptotic distribution

As seen in our simulation experiments, the outcome of the variable addition tests, generally poor, is extremely sensible to the choice of instruments, i.e. to the specification of the marginal (reduced form) model. Some of these results may be explained by the fact that the limiting distribution of the test statistic is nonstandard because $I(1)$ variables are involved. Therefore, in this subsection, we analyse the asymptotic distribution of the Hausman test statistic **H**. The asymptotic distribution of the other two statistics **N** and

W follows from their relation to **H** as can be deduced from (4.15)-(4.17). We use the framework developed in Park and Phillips (1988, 1989) to *sketch* the asymptotic derivations, instead of giving rigorous proofs of the results.

The properties of the test statistic **H** depend crucially on the residuals $\hat{v}_t$. Their population equivalent is defined by

$$v_t = \begin{cases} x_{2t} - E[x_{2t}|z_t] & = & x_{2t} - \pi' z_t & \text{(Case A)} \\ \Delta x_{2t} - E[\Delta x_{2t}|z_t] & = & \Delta x_{2t} - \pi' z_t, & \text{(Case B)} \end{cases}$$

where z_t denotes the vector of instruments. Because Δx_{2t} is stationary, v_t will also be stationary in Case B. In addition, v_t is stationary if there exists a stationary linear combination of x_{2t} and z_t, e.g. if z_t contains x_{2t-1} (sets 1_A and 4_A), or y_{t-1} (set 3_A). The limiting null distribution of **H** in all these cases will be $\chi^2(k_2)$, so that size distortions have to be explained by the use of invalid instruments or by poor asymptotic approximations for particular parameter settings. Therefore we shall concentrate on the case where v_t is non-stationary (set 2_A).

Let us first express the test statistic in matrix notation. We take the convention that variables without subscript t denote the corresponding data matrices or vectors with T rows of observations:

- X is the matrix of regressors of the structural error correction model with dimension $T \times k$
- Z is the matrix of instruments

We further define the projection matrices $P_X = X(X'X)^{-1}X'$, $M_X = I - P_X$ and define P_Z and M_Z analogously. The marginal model is

$$x_2 = Z\pi + v \tag{4.25}$$

so that the OLS residuals are given by $\hat{v} = M_Z x_2$. The augmented structural model is

$$\Delta y = X\beta + \hat{v}\delta + u_1 \tag{4.26}$$

which leads to the OLS estimator of δ :

$$\begin{aligned}\hat{\delta} &= (\hat{v}'M_X\hat{v})^{-1}\hat{v}'M_X\Delta y \\ &= (\hat{v}'M_X\hat{v})^{-1}\hat{v}'M_X u \qquad (4.27)\end{aligned}$$

where the second equality holds only under H_0. The estimated variance of $\hat{\delta}$ is $\hat{\sigma}^2(\hat{v}'M_X\hat{v})^{-1}$, where

$$\hat{\sigma}^2 = \frac{URSS}{T} = \frac{\Delta y' M_{Xv}\Delta y}{T} \qquad (4.28)$$

and $M_X v$, is defined analogously to M_X, with X replaced by $[X \vdots \hat{v}]$.

The Hausman test may now be expressed as

$$\begin{aligned}\mathbf{H} &= \frac{\hat{\delta}(\hat{v}'M_X\hat{v})\hat{\delta}}{\hat{\sigma}^2} \\ &= \frac{u_1'M_X\hat{v}(\hat{v}'M_X\hat{v})^{-1}\hat{v}'M_X u_1}{\hat{\sigma}^2} \qquad (4.29)\end{aligned}$$

where the second equality again holds under the null hypothesis. We state without proof that plim $\hat{\sigma}^2 = \sigma_{11}$ as $T \to \infty$, which will hold if the model is well specified and hence u_{1t} is a white noise error with variance σ_{11}.

We assume that z_t (possibly after a non-singular transformation) may be partitioned as (z_{1t}, z_{2t}), where z_{1t} contains stationary components, whereas z_{2t} is an integrated but non-cointegrated vector. To derive the asymptotic null distribution of the numerator of (4.29), we assume that a multivariate invariance principle[9] holds for $w_t = (u_{1t}, \Delta x_{2t}, \Delta z_{2t})' = (u_{1t}, w_{2t})'$, i.e. we

[9] For this purpose, we basically use two major results which form the basis of the asymptotic theory for regression with integrated processes. In particular, we first need to use a sequence of random variables whose convergence is ensured by a suitable transformation since as such, ordinary central limit theorems do not apply in the case of $I(1)$ processes. This is done by transforming the sequence $\{w_t\}_1^T$ so that each element of the transformed sequence lies in the space of all real valued functions on $[0,1]$. This is achieved by using the concentrated series $w_t^* = \frac{1}{\sqrt{T}}\sum_{t=1}^{[rT]} w_t$ where $[rT]$ denotes the integer part of the corresponding argument so that the original axis -of 1 to T- is concentrated to the closed interval $[0,1]$, indexing the observations by r. For example, if $T = 100$, then w_{50} will be indexed by $r \in [.50, .51]$. Once this is achieved, and under some weak technical assumptions on the temporal behaviour of the original process we can use the so-called *invariance*

assume that, as $T \to \infty$,

$$\frac{1}{\sqrt{T}} \sum_{t=1}^{[rT]} w_t \Rightarrow W(r) = \begin{bmatrix} U(r) \\ X(r) \\ Z(r) \end{bmatrix} \tag{4.30}$$

where "$\Rightarrow$ " denotes weak convergence of the associated probability measure and $W(r)$ is a $(2+\cdot m)$ dimensional vector Brownian motion on [0,1], where m is the dimension of z_t. The structure of the covariance matrix of this Brownian motion is crucial for the limiting distribution of **H**. The variance of $U(r)$ will simply be σ_{11}, i.e. the variance of u_{1t}. The covariance of $U(r)$ and $W_2(r) = (X(r), Z(r)')'$ is given by (see Park and Phillips, 1988)

$$\Sigma_{uw2} = E[u_{1t}w_{2t}] + \sum_{t=2}^{\infty} E[u_{1t}w_{21}] + \sum_{t=2}^{\infty} E[u_{11}w_{2t}] \tag{4.31}$$

Under H_0, both x_{2t} and z_t are predetermined, which makes the first two terms zero. The last term is a little more difficult; however, as long as the marginal models for x_{2t} and z_t do not contain the error correction term, it can be shown that the third term will also be zero, see Johansen (1992a). Note that this is not the case if $\mu \neq 0$ in (4.21). Therefore U and $W_2(r)$ will be uncorrelated only if $\mu = 0$.

Using the fact that $\hat{v}$ contains the residuals of a regression of x_2 on the instruments Z, i.e. $\hat{v} = M_Z x_2$, we have

$$\begin{aligned} \hat{v}' M_X \hat{v} &= x_2' M_Z M_X M_Z x_2 \\ &= x_2'(I - P_Z) M_X (I - P_Z) x_2 \\ &= x_2' P_Z (I - P_X) P_Z x_2 \end{aligned} \tag{4.32}$$

---

principle -see the results in Billingsley (1968), Phillips and Durlauf (1986)- which states that the transformed sequence weakly converges, as $T \to \infty$, to a scalar unit variance Brownian motion, in the case where w_t is a scalar process series, or to a vector Brownian motion in the case of a vector process w_t. The remaining results are obtained using the continuous mapping theorem (see Billingsley, 1968) which states that if $f(.)$ is a continuous function on the space of all real valued functions on $[0,1]$ and if $w_t^* \Rightarrow W(r)$ then $f(w_t^*) \Rightarrow f(W(r))$. The detailed theory is presented in Park and Phillips (1988, 1989), Phillips and Durlauf (1986) and excellent presentations are provided by Dolado(1991), Boswijk (1992b), Banerjee and Hendry (1992).

where the last equality follows from $M_X x_2 = 0$. If the components of Z are not cointegrated with X, so that $M_X Z$ is not stationary, we also have

$$T^{-2}\hat{v}'M_X\hat{v} \Rightarrow \int_0^1 V(r)^2 dr \tag{4.33}$$

where $V(r)$ is some process defined on [0,1]. The precise definition of $V(r)$ depends on whether or not x_1 is included in Z but the important thing to note is that $V(r)$ is a function of $X_2(r)$ and $Z(r)$ alone (hence independent of $U(r)$ if $\mu = 0$). We also can prove

$$T^{-1}\hat{v}'M_X u \Rightarrow \int_0^1 V(r)dU(r) \tag{4.34}$$

still under H_0. Combining (4.29), (4.33) and (4.34) gives the limiting distribution of $\mathbf{H}$:

$$\mathbf{H} \Rightarrow \frac{1}{\sigma_{11}} \int_0^1 dU(r)V(r)[\int_0^1 V(r)^2 dr]^{-1} \int_0^1 V(r)dU(r) \tag{4.35}$$

This type of function of Brownian motion is very common with test statistics (Wald, LM, LR) which are χ^2 for stationary models, see Park and Phillips (1989). In general, the distribution of (4.35) is non-standard and depends on nuisance parameters. However, if $U(r)$ and $V(r)$ are independent, we may condition expression (4.35) on $V(r)$ which results in a $\chi^2(1)$ distribution. Because this distribution is independent of $V(r)$, we have :

$$\mathbf{H} \;\Rightarrow\; \chi^2(1) \tag{4.36}$$

Similarly, if we test the joint weak exogeneity of a vector of h variables, then a $\chi^2(h)$ distribution will be obtained. The above results hinge however on two important assumptions.

1. Firstly, it requires that u_t be serially uncorrelated;

2. Secondly, the covariance $\sum_{t=2}^{\infty} E[u_{11}w_{2t}]$ should be zero.

As noted above, this will fail if x_{2t} or z_t are "error correcting" ($\mu \neq 0$), and also if z_t contains lags of y_t but is nonstationary.

One could naturally refine and generalise these asymptotics but this goes outside the scope of this work. What these simple results indicate is that there are potential sources of deviation from the standard χ^2 limiting distribution. In particular, many empirical studies use lagged values (levels) of all variables included in the structural model so that this may induce important distortion as pointed out. Also, the serially uncorrelatedness of u_t is essential, once again indicating the importance of correct dynamic specification before inference on exogeneity is conducted. Finally, the presence of the cointegrating vector in the marginal model ($\mu \neq 0$) should be carefully investigated seen the potential implications in terms of the limiting distribution.

So far, we have considered the behaviour of orthogonality tests in the presence of cointegrated variables. This has been analysed in the framework of a structural error correction model, but the basic results are applicable to different parametrisation in dynamic models with cointegrated variables. However, the particular model considered has proven its empirical relevance as shown by the growing popularity of single equation ECMs. Our analysis provides results which point out the potential problems that one can encounter by using orthogonality tests with nonstationary time series. Concerning the small sample behaviour, the limited evidence provided by our small scale experiment shows the importance of the signal to noise ratio[10] and of the instrument set on the small sample power of orthogonality tests. Although appealing due to their simplicity of application, (one simply runs two successive regressions using a standard computer package or even preprogrammed routines), orthogonality tests have some drawbacks in the presence of cointegrated variables which range from their poor behaviour in small sample, their potential non standard limiting distribution to the fact that the implicit null hypothesis is usually not sufficient for weak exogeneity in cointegrating systems.

[10]This particular point is worth further investigation and this both with stationary and nonstationary variables as the underlying problem is more related to an expanded regression framework than to the nonstationarity *per se*.

4.4 Testing for Weak Exogeneity in ECMs when the Short Run Dynamic Parameters are Parameters of Interest

When both short and long run parameters are parameters of interest in a structural error correction model, we have pointed out that weak exogeneity nessecitates both the condition of orthogonality and the condition that the "potential" endogenous r.h.s. variable is not error correcting. We thus face the problem that, if we want to build a parametric (weak) exogeneity test in this case, we have to consider the two points in order to assess or reject weak exogeneity.

The first simple solution is to adopt a sequential approach. First test for weak exogeneity with respect to the long run parameter among the lines proposed by Boswijk (1991b) or Johansen (1992a). If the null hypothesis (i.e. absence of the cointegrating relationship in the marginal model) is not rejected, the orthogonality condition can be tested by means of an appropriate orthogonality test. However, this sequential procedure suffers from the problem that if the absence of the cointegrating vector in the marginal model is "wrongly accepted" (i.e. weak exogeneity with respect to the long run parameters is not rejected) then the orthogonality test will be based on a misspecified model within which the behaviour of orthogonality tests has been found in the previous section to suffer from several defaults. Therefore, one might want to test these two separate hypotheses jointly. Such a procedure would be well suited for testing for weak exogeneity in a single equation structural ECM when both the short and long run coefficients are parameters of interest. In this section, we shall point out that such a procedure is easily obtained, within a limited information set-up, by extending the approach of Boswijk (1991b).

In the stationary framework, none of the incomplete model representations emerges as significantly more appealing. In the case of a cointegrated system, however, the implications in terms of model representation are more clear-cut.

Let us come back to a notation similar to that used in Chapter 3. We assume that:

- we have a $k \times 1$ vector time series $\{x_t\}$. We partition x_t as $x_t'=(y_{1t}, y_{2t}, z_t')$. y_{1t} and y_{2t} are scalar time series and z_t is a $(k-2)\times 1$ vector.
- all k components of x_t are $I(1)$ and jointly generated by a cointegrated VAR(p) model, with the cointegrating matrix α of rank 1:

$$\Delta x_t = \sum_{i=1}^{p-1} A_i \Delta x_{t-i} + \gamma\alpha' x_{t-1} + \varepsilon_t \tag{4.37}$$

  γ is the $k \times 1$ vector of loadings. ε_t is a $k \times 1$ vector process identically, independently and normally distributed with zero mean vector and p.d.s. covariance matrix Σ.
- we have $k-2$ predetermined variables in z_t which are not error correcting
- we have two potential endogenous variables in the system, given by $y_t = (y_{1t}, y_{2t})'$

We also consider a *single equation ECM in structural form* for the endogenous variable y_{1t}

$$\Delta y_{1t} = \lambda\alpha' x_{t-1} + \sum_{i=1}^{p-1} \Phi_i \Delta y_{1t-i} + \sum_{i=0}^{p-1} B_i w_{t-i} + v_t \tag{4.38}$$

where $w_t = (\Delta y_{2t}, \Delta z_t')$. λ is the structural error correction coefficient. $\Phi = (\Phi_1, \ldots, \Phi_{p-1}, B_0', \ldots, B_{p-1}')$ denotes the structural short run dynamic coefficients. The parameters of interest ψ are both the short and long run parameters of (4.38): ψ = vec (Φ, α, λ). y_{2t} is the potential endogenous variable and z_t is predetermined and also weakly exogenous for (Φ, α'), e.g. $\gamma_3 = 0$ and $\mathrm{E}(v_t z_{t-s}) = 0$, $\forall s$.

Note that the restrictive nature of this simple framework will be mentioned in the conclusions of this chapter. In any case, the generalisation to tests for weak exogeneity of more than one potential endogenous variable is straightforward although, in that case, the identification problems has to be solved, following for example Boswijk (1991b). A typical empirical example is given by aggregate import models where both imports and prices are potential endogenous variables, but where, often, only one cointegrating relationship is found (see Urbain, 1990, 1992a). Notice that from (4.38) we can obtain a consistent and efficient estimate (see Boswijk, 1991a, 1992a,b) of the cointegrating vector α, that we denote by $\hat{\alpha}$. Our interest lies in the following simple question:

Is y_2 weakly exogenous for ψ ?

In order to derive a simple variable addition test for this hypothesis, we have to complete the structural ECM by a marginal model for y_{2t}. According to the traditional limited information framework, and given the implications in terms of model representation of the GRT, the most natural way to complete the structural ECM of y_{1t} is by adding a *reduced form ECM* for y_{2t}, where the cointegrating vector ($\hat{\alpha}$) is obtained from the structural ECM (4.38). Given (4.37), such a marginal reduced form ECM reads as

$$\Delta y_{2t} = \sum_{i=1}^{p-1} A_{2,i}\Delta x_{t-i} + \gamma_2\hat{\alpha}' x_{t-1} + \varepsilon_{2t} \tag{4.39}$$

where the $A_{2,i}$'s, γ_2 and ε_{2t} stem respectively from the partitioning of the A_i's , γ and ε_t (in (4.37)) accordingly to the partitioning of x_t' in $(y_{1t}\ y_{2t}\ z_t')$.

This *limited information* set-up presents important advantages with respect to the specification of the marginal model. Since cointegration holds by assumption, it enables us to rely on a marginal reduced form (ECM) model which, hopefully, will be less subject to misspecification than an empirically derived marginal model.

Weak exogeneity of y_{2t} for the long run parameters implies the parametric restriction $\gamma_2 = 0$. In order to test this hypothesis, we can follow Johansen (1992a) or Boswijk (1991b) and test for the significance of the (estimated) cointegrating vector in the marginal model, i.e. test for $\gamma_2 = 0$. If $\gamma_2 = 0$, weak exogeneity of y_{2t} for both long and short run parameters only requires the usual orthogonality condition $\text{cov}(v_t, \varepsilon_{2t}) = 0$ which can be easily tested using any of the orthogonality mentioned above.

When short and long run parameters of (4.38) are considered as parameters of interest, then it necessitates $\gamma_2 = 0$ as well as $\text{cov}(v_t, \varepsilon_{2t}) = 0$, which is the usual orthogonality condition, where v_t is the disturbance of the single equation structural ECM (4.38). As pointed out earlier in this chapter, for the purpose of testing the orthogonality hypothesis, two potential theoretically equivalent model representations can be used. However, in the case of an ECM in structural form, representation I (Table 4.1) appears to be more appealing since it allows us to test *jointly* for $\gamma_2 = 0$ and $\mathrm{cov}(v_t, \varepsilon_{2t}) = 0$ by fitting a marginal reduced form ECM such as (4.39). Provided that the structural model is not subject to misspecification or serial correlation we can add, under model representation I of Table 4.1, the fitted residual ($\hat{v}_t$) of the structural equation (4.38) to (4.39). This results in the artificially *augmented* marginal ECM

$$\Delta y_{2t} = \sum_{i=1}^{p-1} A_{2,i}\Delta x_{t-i} + \gamma_2\hat{\alpha}' x_{t-1} + \theta\hat{v}_t + \varepsilon_{2t} \tag{4.40}$$

The null hypothesis of exogeneity of y_{2t} for the parameters of interest ψ is given by

$$H_0 : \gamma_2 = 0\,; \;\; \theta = 0 \tag{4.41}$$

When one of these two parameters is nonzero, weak exogeneity w.r.t. ψ fails. Since all terms in (4.40) are stationary under the assumption of cointegration, this joint hypothesis can be tested by means of omitted variable tests using standard limiting distribution (in χ^2 or F-form).

What happens if we reject the null hypothesis?

The potential problem we encounter depends on which of the two separate hypotheses is rejected or whether they are both jointly rejected. This is a typical multiple decision problem which arises in multiple hypothesis testing. We shall not discuss this here in detail and refer the interested reader to Savin (1984) for a theoretical discussion. In the case where the cointegrating vector does appear in the marginal model (e.g. $\gamma_2 \neq 0$), inference on the cointegrating vector components in (4.38) is complicated by nuisance parameter dependency. Traditional χ^2 statistics cannot be used (see Johansen, 1992a), so that higher dimension systems are eventually required unless the investigator is willing to simulate the limiting behaviour of the test statistics as proposed in Johansen (1992a) or Boswijk (1991b).[11] On the other hand if y_{2t} is not error correcting but orthogonality fails, single equation OLS may yield inefficient or inconsistent estimates of the parameters of interest of (4.38) which include the short run structural dynamic parameters.

[11] Notice that, recently, Boswijk (1992a,b) has proposed two-(or three) step approaches allowing to conduct inference using χ^2 statistics in structural ECMs, even when weak exogeneity does not hold.

4.5 Conclusions

Weak exogeneity is essential for an efficient use of conditional representation of cointegrated systems and therefore necessitates the use of testing procedures if we want to avoid the specification and estimation of full systems. Orthogonality tests, in addition to not being sufficient to assess weak exogeneity in ECMs, are not extremely appealing for a number of reasons including their poor small sample behaviour (as in the stationary case), their potential nonstandard limiting distribution as well as their sensitivity to the parameter values. On the other hand, the procedures proposed by Johansen (1992a) and Boswijk (1991b) address the problem well w.r.t. the long run coefficients and are relatively easily extended to joint tests for both orthogonality and absence of cointegrating vector in the marginal model.

With respect to the model representation, contrary to the stationary case, the restrictions that have to be tested for assessing weak exogeneity favour the representation of the incomplete SEM advocated by, for example, Richard (1984) and Steel (1987b). In particular, a *Limited Information* set up is in accordance with the Granger Representation Theorem, and therefore provides an attractive framework for addressing this problem.

It is also important to note that the framework we have used masks an essential device that any attempt to test for weak exogeneity should be aware of. It is the reliance on the assumption of a single cointegrating vector. If there is more than one cointegrating vector in the system, then a single equation will only identify a linear combination of them. Whether it is *that linear combination* which corresponds to the economically meaningful long run relation of interest is impossible to say ex-ante, but is not unlikely to occur in practice (see the discussion in Clements and Mizon, 1991). If it is the case, we can proceed along the lines presented here without generalisation. We therefore think that it is important to rely on a framework which addresses this by considering the identification of the long run relationships *explicitly* as in the approach proposed by Boswijk (1991a). When there is more than one cointegrating relationship of interest the fundamentals of our analysis remain valid, but the latter must be modified in order to consider the problem within a sub-system of higher dimension.

Chapter 5

Empirical Analysis: The Case of Aggregate Imports

In this chapter we shall now illustrate the modelling of cointegrated systems by studying empirically Belgian's aggregate imports. Contrary to topics such as money demand or consumption functions, the problem of the dynamic modelling of trade flows has not often been used as explanatory field for recent econometric techniques and modelling approaches. Nevertheless, we think that a refinement in the dynamic specification of trade flows may lead to a substantially different analysis of price and income effects in international trade.

5.1 Background

The econometric modelling of aggregate trade flows has a long history in the economic and applied econometric literature. Starting in the late 50's with Cheng (1959), a number of important surveys have been published including Leamer and Stern (1970), Magee (1975), Stern et al. (1976) and more recently Goldstein and Khan's (1985) often quoted contribution to the Handbook of International Economics. One of the reasons for this important amount of published studies is certainly the fact that the underlying (economic) theoretical framework for the determination of prices and trade volumes is usually remarkably simple and familiar from standard consumer demand or production theory. On the other hand, the effectiveness of international trade policies is dependent on the size of price and activity effects on trade flows so that policy makers show important interest in reliable estimates of these parameters. From an econometric point of view it has been however surprising, as pointed out by Mayes (1981), that among all fields of applied econometrics, researchers dealing with the modelling of trade flows have been quite reluctant to integrate the recent advances in time series econometrics. Exceptions include the contributions by Gagnon (1989), Husted and Kollintzas (1984, 1987), Kollintzas and Husted (1984), Clarida (1991) who develop theoretical based dynamic stochastic models for aggregate imports, while Urbain (1988,1990, 1992a) and Asseery and Peel (1991) use cointegration theory for the empirical modelling of aggregate imports.

This chapter extends the results obtained in Urbain (1990, 1992a), which were based on annual data for the period 1953-1985. The structure of the chapter is as follows. In the first section, we present some of the existing work on aggregate imports and focus on its potential defaults. The arguments presented here partly overlap some of those reported in Urbain (1990, 1992a). Empirical results for Belgium are reported in the second section where the univariate properties of the data are first analysed. Both full and partial systems are used for the derivation of a dynamic empirical model for aggregate imports. The question of exogeneity of import prices is also investigated in this framework.

5.1.1 The imperfect substitutes model for aggregate trade flows

Although the nature of the goods being traded, the end-use and the geographical orientation are factors which are important in modelling disaggregated trade flows, we restrict attention to total aggregate trade flows so that the basic underlying economic theory can be remarkably simple.

Traditionally (see Leamer and Stern, 1970, Goldstein and Khan, 1985) the basic question is whether imports (and exports) are, or are not, perfect substitutes for domestically produced goods. Since two-way trade is usually observed, i.e. imports, exports, domestic production and intra-industry trade, a perfect substitute model is clearly ruled out as a potential background for modelling international aggregate (but even disaggregated) trade flows.

The imperfect substitutes model (for details, see Goldstein and Khan, 1985) is briefly described bellow. For simplicity, we consider a country's i imports from the rest of the world:

$$\begin{aligned} M_i^D &= f(Y_i, PM_i, PD_i) \\ X_i^D &= g(Y^*e, PX_i, P^*e), \\ M_i^S &= h(PM^*(1+S^*), P^*), \\ X_i^S &= j(PX_i(1+S_i), P_i*), \\ PM_i &= PX^*(1+T_i)e \\ PM^* &= PX_i(1+T^*)/e \\ M_i^D &= M_i^S e \\ X_i^D &= X_i^S \end{aligned} \tag{5.1}$$

where f_1 and $f_3 > 0$, $f_2 < 0$, g_1 and $f_3 > 0$, $g_2 < 0$, $h_1 > 0$, $h_2 < 0$, $j_1 > 0$, $j_2 < 0$. M_i^D is the quantity of total imports demanded in country i, M_i^S the quantity of imports supplied to country i by the rest of the world, Y_i (resp. Y^*) the nominal income from country i (resp. the rest of the world), PM_i (resp. PM^*), the price (in domestic currency) paid by the importers, i.e. the import price, X_i^D country's i export demanded by the rest of the world, X_i^S country's i export supply to the rest of the world, PX_i (resp. PX^*) the price in domestic currency received by exporters in country i (resp, rest of the world), e the exchange rate expressed in the country's i unit, (T_i) (resp. T^*) proportional tariffs, PD_i* (resp. PS^*) the domestic price, S (resp. S^*) subsidy rates applied to imports and exports.

For notational convenience we use a simple, hence unrealistic, two-country framework, which implies symmetry between demand for imports and export supply. The latter symmetry does naturally not hold anymore when "the rest of the world" is replaced by a number of n foreign countries. For an extensive presentation of the above imperfect substitutes model and the discussion of the underlying assumptions, we refer the interested reader to Goldstein and Khan (1985). The equilibrium characteristics underlying the imperfect substitutes model are, in any case, worth a brief remark. They imply that prices move to equate demand and supply in each time period. This may however appear as inconsistent with the recent view of imperfect competitive markets,[1] where adjusting prices is not costless, especially when we allow for output costs, costs arising from changing inventories,... There are several alternatives to deal with the apparent inadequacy of models developed in this framework and the economic reality. Among these we can eventually keep the imperfect substitutes model as the basic framework, and retain the non-market clearing property of the model by including non-price rationing variables (as for example capacity utilization rate, see Italianer, 1987) in both the demand and supply equations, so that price changes are not the only sources of market clearing. We can also assume that observed trade quantities are in disequilibrium and model the adjustment scheme resulting in dynamic models for aggregate imports. An alternative solution is to consider a different theoretical framework allowing for imperfect competition and derive the optimal behaviour of a representative trader in an intertemporal framework. This last solution is chosen among others by Gagnon (1988).

In practice, however, most of the existing empirical models base their specification on (5.1), or on a modification thereof, as clearly pointed out in the review of Goldstein and Khan (1985). Under infinite supply elasticity, it is argued that (5.1) can be estimated by single equation estimation techniques. Notice that this condition is necessary from an economic point of view, but not from a statistical viewpoint which only requires weak exogeneity of the r.h.s. variables.

5.1.2 Econometric issues in trade modelling

Orcutt's critique

Since the early empirical studies of trade relations, most researchers have been aware of the statistical problems underpinning the econometric modelling of aggregate trade flows. This is mainly due to the seminal paper by Orcutt (1950) who stressed the potential problems underlying single equation estimation of models like (5.1). Consequently, subsequent researchers have tried to "correct" their estimates so that these could not be subject to these critics, but without really questioning the modelling framework

[1] A recent non-technical survey of the main ideas that have been recently developed in international trade is presented in Oliviera (1990).

on which they were based, exceptions made of the work by Husted and Kollintzas (1984, 1987), Kollintzas and Husted (1984) and more recently Gagnon (1988), Kholi (1990), Marquez (1991), Clarida (1991), Marquez (1991). This has become the *Orcuttisation* of empirical trade modelling, a label given by Magee (1975). Note that more than forty years after Orcutt's contribution, the latter is still motivating attempts to model imports and exports in a different framework (see for example Marquez, 1991). The empirical relevance of the point raised by Orcutt (1950) has been extensively analysed over the last decades with more or less fortune. An account of these is given in Magee (1975) and Goldstein and Khan (1985). We shall not discuss these in details. We focus instead on some points which, we think, can be naturally addressed in an empirical modelling framework which take into account both the time series properties of the data as well as the joint dependent character of probably most of the variables involved in the specification of import demand equations.

The *simultaneity bias* is central to Orcutt's critique and stems from the idea that single estimates are biased due to the simultaneity of prices and quantities, i.e. to the lack of identification of the true demand equation. The idea is that simultaneity can occur between quantities and prices so that a single equation specification implies estimates which are weighted averages of demand and supply elasticities. To overcome this problem one usually assumes an infinite supply elasticity. Tentative work based on simultaneous models has been pursued, without convincing empirical success (see Goldstein and Khan, 1985). As pointed out by Haynes and Stone (1985), most supply studies use supply-quantity models where the quantity supplied is a function of both present and past prices, resulting in the well known lack of identification. When the firms act in a uncertain world, they will probably pursue a pricing policy which is based on past market conditions so that the supply model becomes more naturally a price-supply equation. Even in this case, exogeneity of the prices is open to empirical analysis if the derivation of price and income effects in international trade has be to investigated. Recently, Marquez (1991) has taken a different point of view by estimating a complete demand system, so that simultaneity is taken into account.

Not unrelated to the previous point is the problem of *instability* in aggregate import demand studies, a point also implicit in Orcutt (1950). The fact that structural change is likely to affect estimated aggregate import functions has attracted many empirical researchers. It is argued that when studying a data set covering a large span of years, the economic development process covers regimes which imply different degrees of openness of the various national economies so that instability is natural. In this case the usefulness of elasticities estimates is doubtful and their use for conditional simulation of, for example, devaluation policies or other related economic policies can be questioned.

Another interpretation of the observed instability of trade relations is given by Marquez (1991). His argument is similar to that implicit in the Lucas (1976) critique, namely that single estimation of aggregate import demand models based on equations like (5.1) are likely to display instability since the coefficients of such a model will likely be mixing structural parameters with expectations and adjustment parameters. On this latter point, see also Gagnon (1988). Marquez (1991) therefore derives import demand models based on the development of a theoretical model of optimization behaviour for a representative importer, allowing him to model jointly quantity and prices and, according to the author, derive meaningful parameters. Since the framework used is based on a Rotterdam demand system, the estimated elasticities will vary with the reallocation of expenditures, so that, unless the underlying preferences are of Cobb-Douglas type, a model like (5.1) will display unstable coefficients. Nevertheless, there are some limitations to Marquez' (1991) approach which are likely to alter his conclusions. First of all, the preference independence on which the model is based rules out complementarity. While this may be rationalized in the case of the US imports, it is less suited for the Belgium economy where an important part of total imports is composed of complementary products in the production process. From an econometric point of view several technical points of this paper need also to be raised. In particular, and contrary to recent trends in time series econometrics, the model and parametrisation used by Marquez is a static model, not allowing for any dynamic adjustment mechanisms. On the other hand, as pointed out in Urbain (1988, 1990, 1992a) and Asseery and Peel (1991), almost all variables used in the estimation of aggregate import functions are highly trending variables which are well represented by (co)integrated processes. Not taking this into account leads to an important potential loss of information when the first differences are modelled, as in Marquez (1991). A final potential critique is that he gives an economical explanation to coefficient instability which is often known to be a sign of misspecification (see Anderson and Mizon, 1990).

In any case, we agree with Marquez (1991) that simultaneity between prices and trade volume should be allowed for. Nevertheless, our approach is totally different, and, as much as possible, we avoid using the term *elasticities* in favour of the notions of short and long run "effects" of the various factors on import volume and prices. The question whether estimated elasticities obtained from single equation models should be treated as *autonomous parameters* (in the sense of Aldrich, 1989), as it is implicit in most of the empirical work done on import demand, or not, such as argued by Marquez (1991), is a point which does clearly go far beyond the scope of this chapter. Elasticity is a theoretical concept which is effectively linked, in traditional consumer demand theory, to expenditure shares so that, as a theoretical concept, it varies with the latter shares. On the other hand the question which is also implicit in Marquez' argument is: do we believe in constant or non-constant parameters? In this particular case, we follow the point of view that econometric modeling is aimed to deduce invariant characteristics of the observed world (see the discussion in Hendry and Richard, 1982, 1983, Ericsson, 1992)[2] so that stability of the parameters is an important property which a valid empirical (econometric) model has to satisfy if we want to be able to discuss alternative policy issues. This is naturally related to the notion of super exogeneity which is a useful concept for the elaboration of dynamic macroeconometric models whose purpose is policy analysis. Being able to "extract" invariant relations between imports, prices and income, by allowing the discrimination between short and long run responses is, according to our belief, a strong argument in favour of the associated model.

The dynamic modelling of aggregate imports

Early studies of aggregate trade flows have recognized the need for introducing dynamics into empirical trade models. The basic demand and supply equations are interpreted as equilibrium relationships, to which quantities and prices do not necessarily adjust at each period of time. The presence of adjustment costs, incomplete information, delivery lags, decision lags in the real world are acknowledged as the main sources of deviation from the equilibrium so that the gradual adjustment to the latter has to be modelled.

[2]This does not rule out that we might assume a time varying model, but in that case, the constancy becomes that of the associated parametrisation of the evolution of the time varying coefficient.

The justification for lagged behaviour has been studied since the 60's (see Junz and Rhomberg, 1965) and has taken significant space in any survey of trade modelling (see Magee, 1975; Goldstein and Khan, 1985). Note that it has also been justified from a theoretical point of view by Kollintzas and Husted (1984).

Given the variety of dynamic specifications existing in the literature (see Hendry et al., 1984) it is hardly surprising that no consensus exists about the way dynamic has to be introduced. Four specifications have been popular over the years: (*i*) the simple Partial Adjustment Models (PAM), (*ii*) distributed lags, (*iii*) Koyck and (*iv*) Almon polynomials (see the survey in Goldstein and Khan, 1985). However, their empirical success varied from study to study and a common characteristic is that the dynamic specification is usually restricted to prices. These are supposed to affect import volume during a long period of time while income or scale effects are often restricted to the current period. In a comparative study of nine popular specifications of aggregate imports, Thursby and Thursby (1984) found out unsatisfactory results for the "traditional specifications".

Another interesting route of research has been followed by a number of authors who, starting with Husted and Kollintzas (1984), found theoretical justifications for the use of unrestricted distributed lag model. Since adjustment is costly, expected future market conditions should influence current trade flows. Assuming rational expectation of the importer, these authors show that optimizing firms set their (intermediate) import schedules so that they equate the expected cost of the last imported unit in any period, and carried forward to the next period, with the expected cost of the last unit to be imported the next period (see Kollintzas and Husted, 1984). They end up with a distributed lag model whose parameters are structural in the sense of Sargent (1981), and therefore not subject to the Lucas critique. Husted and Kollintzas (1984) provide some empirical support for their model for a selected group of US commodities. The approach has been further developed in Gagnon (1988) who derives a model for international trade flows based on a representative importer who maximizes a discounted sum of expected future operating profits. Using bilateral data, the author argues strongly in favour of his specification as the distributed lag models which result from his analysis have a clear structural interpretation. However, and this is common to almost all the existing work in this field, except the recent paper by Clarida (1991), the treatment of the trending character of the data in Gagnon

(1988) seems open to criticism. The data are filtered with exponential deterministic trends and traditional Granger causality tests are carried out in order to specify models for the forcing variables (which are here among other the prices). As the stochastic nature of the trend is not allowed for, cointegration is not investigated which can be the source of important drawbacks in the specification of the model and especially in the outcome of the causality tests (see for example the discussion in Stock and Watson, 1989).

The trending character of the series

As just pointed out, in the enormous literature on import modelling, almost none of these studies have recognized the potential stochastic nonstationary character of the individual time series under study. When some trending character is taken into account, it is only by allowing the income variables to be decomposed into secular income (so-called trend income) and cyclical income (deviation from trend income), see Goldstein and Khan (1985). This implies a trend stationary assumption for real domestic gross product, an hypothesis which is not accepted overall but which is convenient from an economic point of view as it is likely that imports react differently to a secular or a cyclical change in aggregate demand. Notice that the trending character of prices is not taken into account either.

According to our knowledge, the first attempts to model aggregate imports by means of cointegrated systems are reported in Urbain (1988, 1990, 1992a) and also in Asseery and Peel (1991), Clarida (1991). Using annual data for Belgium and the Netherlands and using both single equation dynamic specification analysis as well as the Johansen procedure (see Urbain, 1988, 1990), we were able to find significant support for the use of an ECM specification. Similar evidence is provided by Asseery and Peel (1991) for quarterly data, although the latter study only considers reduced form ECMs where no contemporaneous differences appear in the model, so that their "short run" coefficients are hardly interpretable. Notice that "ex-post" rationalization of the use of ECM can eventually be found by quoting Gagnon (1988):

> *"The real world almost certainly has mechanisms by which continuing trade imbalances can influence relative prices and aggregate demand and thus damp themselves down over time."*

Recently, Clarida (1991) used a two-goods version of Hall's (1978) representative agent permanent income model to derive a structural import demand model whose parameters are estimated using Phillips and Loretan's (1991) nonlinear least squares estimator. Again, the results are extremely interesting for the US economy but hardly sustainable for small open European countries for which imports are mostly constituted of intermediate products and durables goods. Further work in this direction is however desirable, and clearly constitutes a fruitful route for further research.

The use of a relative price ratio

It is interesting to note that the use of a relative price ratio, instead of two separate price terms, for capturing the price effects on trade flows has since decades been the rule in almost all empirical studies of aggregate imports. As already pointed out by Murray and Ginman (1976), there are, in any case, economic and econometric arguments that clearly favour the use of two separate price terms. The use of relative price terms is motivated by the usual homogeneity constraint which requires the absence of money illusion. As noted by Magee (1975), this is a rather strong hypothesis that applies to both demanders and suppliers in the domestic markets and is more demanding than just absence of money illusion on the part of either group separately. From an econometric point of view, this zero homogeneity restriction, very often imposed, also conveniently reduces the colinearity that occurs between the price terms in models where only level terms are modelled. However, there are some reasons to believe this assumption to be inappropriate.

Like every aggregate price index, import and domestic prices are weighted averages of prices corresponding to different commodity groups. In a total aggregate context, it will rarely occur that a given commodity will have the same weight in both price indices. If both the domestic and the import price of a good rise by the same amount, the resulting change in the aggregate domestic and import prices will generally not be equal (see the discussion in Murray and Ginman, 1976; Urbain, 1992c). This is particularly important for Belgium where about 65- 70% of imported goods are intermediate and equipment goods not available on the national market. Therefore, the commodities entering the construction of the price indices are likely to differ substantially. The problem is even more important when the modelling of aggregate import is considered in a dynamic framework. Zero homogeneity implies identical dynamic response patterns of import volume to changes in both prices. If economic agents try to expect price changes, they will form their expectations using different information sets for the two prices. Expectations about foreign price changes are likely to be influenced by different economic variables. In particular, exchange rate movements are reflected in import prices. One may thus expect short run domestic price effects to be more important in magnitude or at least to occur more immediately than import price effects. Since many countries import goods which have to enter the production process to replace nonexisting nationally produced goods, the reallocation will often take more time or even be impossible for some small open economies. One solution proposed by Wilson and Tackacs (1979) is to impose homogeneity in the long run only by arguing that the short run patterns may differ, but the overall long run effects are symmetric. Even here, one can argue that if the short run behaviour differs radically, the long run effects will not necessarily be equal, so that the price homogeneity has, if possible, to be tested even in the long run before being imposed on the model.

5.2 System versus Partial Approach to the Modelling of Belgian Aggregate Imports

From our previous discussion of the existing work on aggregate import modelling, it appears clear that three important topics are of interest in order to obtain reasonably good estimates of income and price effects:

- the potential simultaneity between prices and import volume which can distort the inference based on single equation import demand models;
- the dynamic specification and the trending character of the series;
- the use of a relative price and the underlying homogeneity assumption.

In the light of the results obtained in the previous chapters of this work, as well as of the empirical results reported in Urbain (1990, 1992a), it appears that this field is an interesting candidate for the illustration of the choice between full versus partial modelling of cointegrated systems. We first follow the proposal of Hendry and Mizon (1990), which consists in deriving a complete SEM for the variables we want to model. A different approach is then followed where some of the variables are considered as potentially exogenous so that their generating models are not explicitly modelled.

5.2.1 Univariate analysis

The data

The variables used in the analysis are those usually entering empirical models of aggregate import demand: import volume, import prices, domestic prices and real income. These are justified by the imperfect substitute model as well as by those derived by Gagnon (1988), Husted and Kollintzas (1984, 1987). Following our previous discussion we do not use relative price but rather two separate price terms. Their symmetry will be tested in due course. The data are quarterly series covering the sample period 1964:02-1990:01. All variables are transformed in natural logarithm. Their time series behaviour, both in level as well as in first differences are depicted in the Figures 5.1-5.7. In the remaining part of the empirical analysis, lowercase letters denote the natural logarithm of the corresponding variables and "Δ_k" defines the kth difference, i.e. $\Delta_k x_t = x_t - x_{t-k}$. Further combinations of the difference operator are sometimes used for notational convenience: $\Delta_i \Delta_j x_t = (x_t - x_{t-j}) - (x_{t-i} - x_{t-i-j})$.

All the empirical calculations have been undertaken with RATS 3.10 and P.C.GIVE 6:01.

For imports, denoted by M, the selected data series is the import volume (quantity deflated by value), the domestic price index (denoted by PD) is proxied by the wholesale price index , import price (PM) is the unit value of imports while for the activity (or demand) variable, we use GDP. As usual in import demand studies, the latter variable is deflated by the domestic price index and denoted by Y. The price and import series are taken from various issues of the I.M.F. financial statistics, while the income variable is taken from various issues of "Cahiers Economiques de Bruxelles".

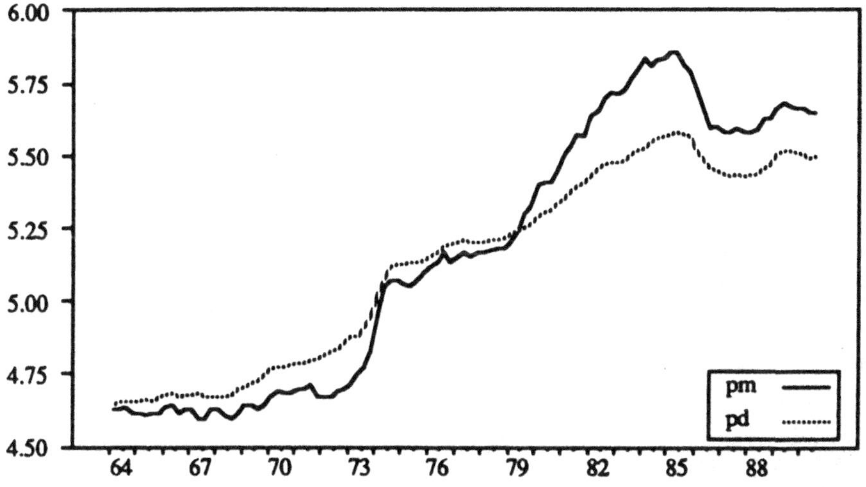

Figure 5.1: Import and domestic prices (in level)

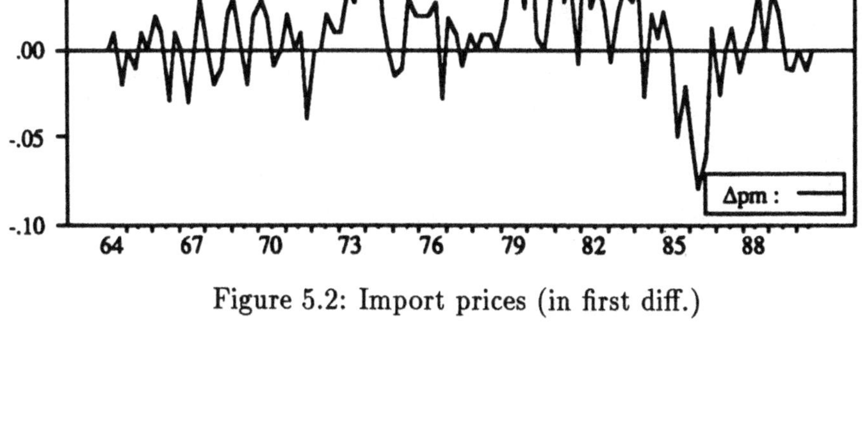

Figure 5.2: Import prices (in first diff.)

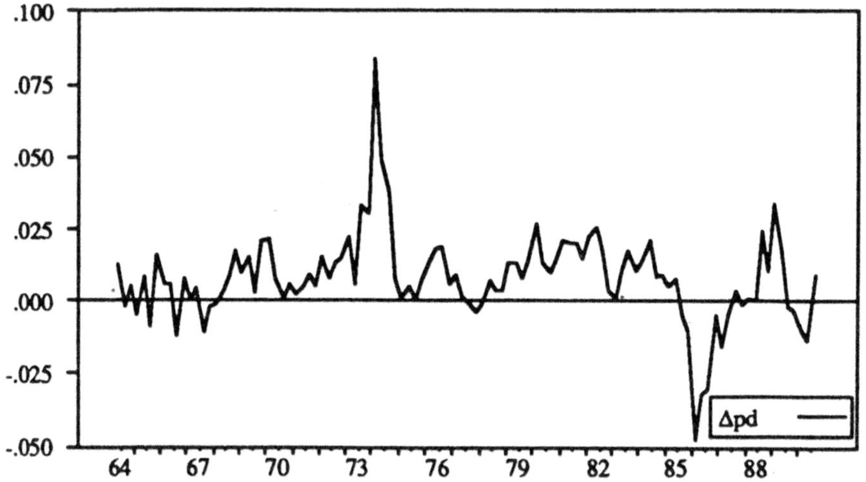

Figure 5.3: Domestic prices (in first diff.)

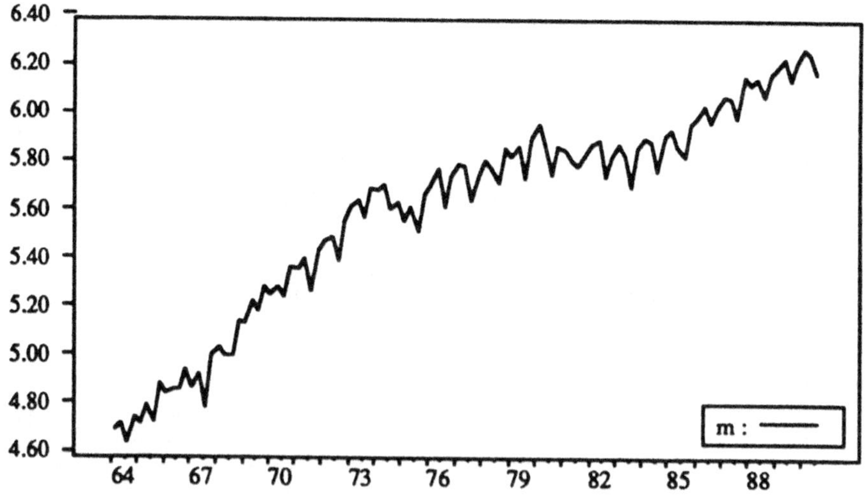

Figure 5.4: Import volume (in level)

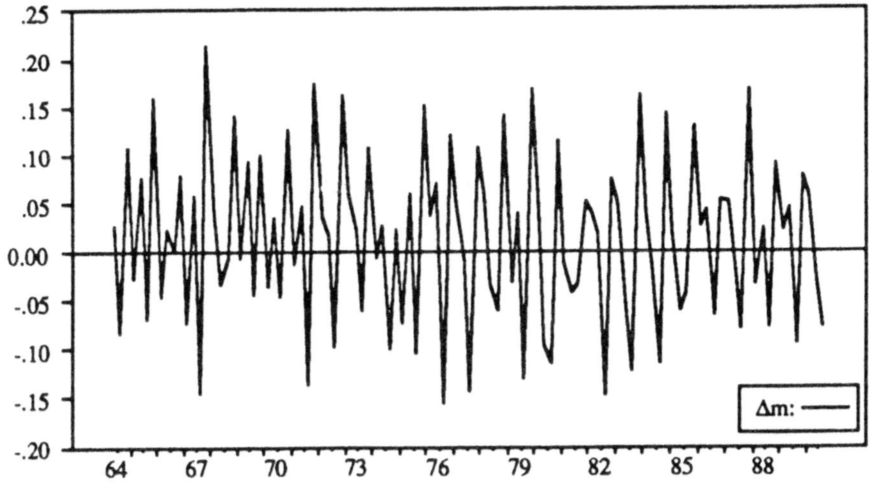

Figure 5.5: Import volume (in first diff.)

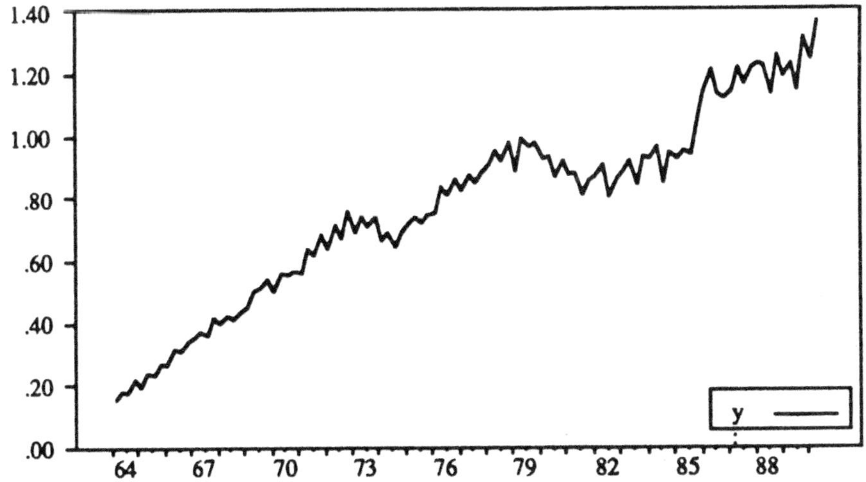

Figure 5.6: Real income (in level)

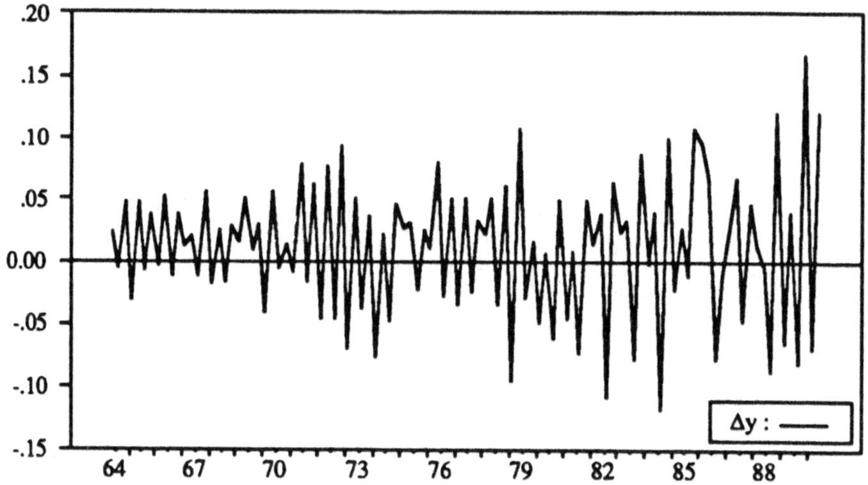

Figure 5.7: Real income (in first diff.)

The problem of seasonality, the order of integration of the series and the effect of the 1973 oil shock.

Since our data are quarterly time series, the problem of seasonality has to be accounted for. In particular, we dispose of a complete set of seasonally unadjusted quarterly time series, so that the danger of bias in favour of the unit root hypothesis, which is often encountered with seasonally adjusted data, is ruled out. On this point we refer to Ghysels and Perron (1991), while for some general discussion of model selection in the presence of seasonality the reader is, for example, referred to Franses (1991b).

In order to allow for the possibility of unit roots at seasonal frequencies, we used the testing strategy described in Hylleberg et al. (1989), hereafter denoted HEGY, generalised by Franses (1991a) for monthly, in an attempt to search for the presence of seasonal unit roots. The HEGY procedure is based on the factorisation of the fourth difference operator $\Delta_4 = (1 - L^4)$:

$$(1 - L^4) = (1 - L^2)(1 + L^2) = (1 - L)(1 + L)(1 + iL)(1 - iL) \qquad (5.2)$$

which allows for four different unit roots. Those corresponding to $+i, -i, -1$ are the seasonal unit roots while that corresponding to 1 is the usual non-seasonal unit root. Note that the practice of using fourth differences of quarterly series is likely to induce overdifferencing when, for example, only one of these four roots is present. In the spirit of traditional Dickey-Fuller tests (Fuller, 1976; Dickey and Fuller, 1979; 1981), the HEGY procedure is based on the following regression model:

$$\Delta_4 x_t = d_t + \pi_1 y_{1t-1} + \pi_2 y_{2t-1} + \pi_3 y_{3t-2} + \pi_4 y_{3t-1} + \sum_{i=1}^{k} \theta_i \Delta_4 x_{t-1} + \varepsilon_t$$

where d_t includes deterministic components. The outcome of the procedure is dependent on these. A constant gives rise to a linear trend, seasonal constants to four different trends while a trend term gives rise to a quadratic trend. We further have

- $y_{1t} = (1 + L + L^2 + L^3)x_t$ which is the x_t process adjusted for the seasonal unit roots of the frequencies $\pi/4, \pi/2$ and $3\pi/4$.
- $y_{2t} = -(1 - L + L^2 - L^3)x_t$ is the process adjusted for the roots at the zero frequency, $\pi/4$ and $\pi/2$ frequency;
- $y_{3t} = -(1 - L^2)x_t$ is the process adjusted for the roots at zero and a half.

The polynomial in the lagged fourth difference ($\sum_{i=1}^{k} \theta_i \Delta_4 x_{t-1}$) allows for higher order processes. One-sided t- tests on the π_i's allow to test for the presence of unit roots at the various frequencies. In particular, π_1 relates to the factor $(1 - L)$, π_2 to the factor $(1 + L)$ while π_3 and π_4 to the factors $(1 - iL)$ and $(1 + iL)$ where $i = \sqrt{-1}$. For the last terms, a joint F-test is conducted.

The length of the lag polynomial was varied from 2 to 6 and, as the results remained almost unchanged for length greater than 3, we report these for the case where a third order lagged polynomial is added. I,S,T respectively denote the presence in the estimated model of an intercept, seasonal dummies and a linear trend. The subscripts "*" and "**" denote significant at respectively 5% and 10% level. Critical values are reported in HEGY.

As shown in Table 5.1 the presence of unit roots at seasonal zero is rejected for all series when seasonal dummies are introduced. The same battery of tests applied to the series filtered by seasonal dummies leads to

Variable	Specification	Π_1	Π_2	Π_3	Π_4	F-test
m	I,S	-1.855	-3.839*	-5.723*	-1.594*	31.33*
	I,T	-2.210	-1.652**	-2.689*	-0.480	3.851*
	I,S,T	-1.962	-3.794*	-5.659*	-1.577**	18.42*
pm	I,S	-0.987	-4.148*	-2.355	-5.905*	21.544*
	I,T	-1.945*	-3.712*	-2.589*	-5.623*	19.777*
	I,S,T	-1.967	-4.02*	-2.501	-6.023*	22.672*
pd	I,S	-1.167	-2.969*	-0.2404	-6.948*	25.257*
	I,T	-1.389	-2.930*	-0.416	-6.918*	24.794*
	I,S,T	-1.434	-2.896*	-0.361	-7.026*	26.418*
y	I,S	-0.714	-1.883	-3.133**	-2.829*	9.605*
	I,T	-2.343	-0.086	-3.077*	-2.460*	8.018*
	I,S,T	-2.216	-1.775	-3.192*	-2.847*	9.764*

Table 5.1: Testing for seasonal unit roots

the rejection of all seasonal factors in our series and is therefore not reported. Given the lack of support for the presence of seasonal unit roots, we shall use centered deterministic seasonal dummies in the subsequent multivariate models.

For each series, the presence of a unit root at the zero frequency is however not rejected (see row 3), i.e. the series seem well characterised by a single unit root. The test on $(1 - L)$ is likely to be less powerful than traditional unit root tests which do not consider the possibility of seasonal unit roots. The order of integration of the series is therefore also investigated using several traditional unit root tests including Augmented Dickey-Fuller (ADF) tests and Phillips-Perron's (1988) modified Dickey-Fuller test. The ADF test considers the following regression model

$$\Delta y_t = \alpha + \mu T_t + \rho y_{t-1} + \sum_{i=1}^{p} \psi_i \Delta y_{t-i} + \varepsilon_t \qquad (5.3)$$

which is estimated by OLS. This model assumes that the univariate time series y_t can be approximated by a finite order autoregressive model (of order

$p+1$). The normalised bias (i.e. $T(\hat{\rho})$), where T is the sample size, or the t-test for $\rho = 0$ is computed. Several assumptions concerning the deterministic components can also be checked using one-sided tests whose critical values are given in Fuller (1976). The limiting distribution is nonstandard and depends on the deterministic components of the univariate series. Critical values are tabulated in Fuller (1976). Phillips-Perron's (1988) modification to the Dickey and Fuller test (based on (5.3) without lagged first differences) is motivated by the fact that the DF and ADF tests are only valid under the assumption of i.i.d. processes. In practice, it may be more realistic to allow for some dependence among the ε_t's. In that case, the asymptotic distribution changes. Phillips (1987) has weakened the i.i.d. assumption by allowing the innovation sequence to be generated by more general dependent processes. The conditions assumed to hold by Phillips allow for serial correlation and heteroscedasticity, but also control the extend of temporal dependence so that, although there may be substantial dependence among recent events, events which are separated by long intervals of time are almost surely independent. Phillips and Perron (1988) have extended Phillips' approach to the case of non zero drift and deterministic trend while Ouliaris et al. (1988) even allow for higher order trend polynomials. The modified t-statistic proposed by Phillips and Perron (1988) in this case is given by:

$$Z_{t_\mu} = (\sum_{t=1}^{T} y_{t-1}^2)^{1/2}(\hat{\alpha} - 1)/s_\ell - (s_\ell^2 - s_\varepsilon^2)T^3/4s_\ell(3D_{xx})^{-1/2}$$

where D_{xx} is the determinant of the regressor cross-product matrix; s_ℓ^2 and s_ε^2 are respectively consistent estimators of σ^2 and σ_ε^2 with $\sigma^2 = \lim E(T^{-1}S_T^2)$ and $\sigma_\varepsilon^2 = \lim T^{-1}\sum_{t=1}^{T} E(\varepsilon_t^2)$. If we assume that ε_t is i.i.d. then $s_\ell^2 = s_\varepsilon^2$. Z_{t_μ} is an adjustment to the traditional t-statistic mentioned before, so that these statistics have the same limiting distribution as the traditional Dickey-Fuller tests, but allow for more general innovation sequences. A consistent estimators of σ_ε^2 is given by $T^{-1}\sum_{t=1}^{T}(y_t - y_{t-1})^2$. For the long run variance, σ_ℓ^2, we use the Newey and West (1987) non-negative estimator:

$$s_\ell^2 = T^{-1}\sum_{t=1}^{T}(y_t - y_{t-1})^2 - 2T^{-1}\sum_{j=1}^{\ell}\frac{1-j}{\ell+1}\sum_{t=j+1}^{(} y_t - y_{t-1})(y_{t-j} - y_{t-j-1})$$

where ℓ is known as the truncation lag. Although theoretical devices do exist for the choice of the lag truncation, it essentially remains an empirical question. Phillips (1987) indicates that ℓ should increase with the sample size. As pointed out in some Monte Carlo simulations, the outcome of the test statistics is sensitive to ℓ (Phillips and Perron, 1988, Kim and Schmidt, 1990). The asymptotic limiting distribution is the same as that of the τ_μ tests tabulated by Fuller (1976).

The results of the unit root tests are reported in Table 5.2.

Variable	τ	τ_μ	Z_τ	Z_{τ_μ}	k
m^a	-1.942	-2.720	-2.647	-2.753	4
y^a	-0.510	- 2.251	-0.661	-2.373	4
pm	-1.296	-1.227	-0.705	- 0.887	2
pd	-1.719	-1.249	-1.251	-0.626	4

Table 5.2: Testing for unit roots at the zero frequency

The superscript "a" denotes that the corresponding variable has been filtered with seasonal dummies in order to ensure that the apparent deterministic seasonality will not affect the outcome of the unit root tests. The price terms are not adjusted as they do not show any sign of seasonality. τ and τ_μ are Augmented Dickey Fuller tests, Z_τ and Z_{τ_μ} are PhillipsPerron non-parametrically corrected version thereof. The subscript μ indicates that a linear deterministic trend was added to the estimated model. k is the lag augmentation used both in the ADF procedure as well as in the construction of a consistent estimates of the long run variance required in the Phillips-Perron tests. In the latter case, this parameter has been chosen by scrutinizing the correlogram of the first differences of the series. The 5% critical values for T=100 are -3.45 when a trend term is present in the estimated model and -2.89 when only a constant is included (see Fuller (1976, Table 8.5)).

The same battery of tests applied to the first differences of the series leads to a clear rejection of the unit root hypothesis in the first differences so that these results are not reported. Our series seem thus well characterised as $I(1)$ processes displaying a single unit root in their autoregressive part. As has been pointed out by many authors (including Schwert, 1989; Schotman and Van Dijk, 1991, Campbell and Perron, 1991) one should however be careful and not draw definitive conclusions from these univariate unit root tests which are known to be particularly fragile in several context. In particular their poor power or substantial size distortions are documented in the case of near-unit root processes (Schwert, 1989), of stationary processes displaying shift in their mean and/or trend function (Perron, 1989; 1990; 1991); or of other violation of the n.i.i.d. framework (see for example Schwert, 1989; Urbain, 1991c).

In particular, we shall pay attention to the constant parameter assumption of the model representation underlying DF-type tests. In any study of trade flows, the 1973 oil shock has to be considered, at least as potential structural break, especially in the price series. We therefore used the testing procedure described in Perron (1989) which allows, under the alternative stationary hypothesis, the univariate series to be stationary around a linear deterministic trend function with structural breaks. We considered the different versions of Perron's tests, implying one-time shifts in the trend function (" innovation outlier assumption") or more gradual changes ("additive outlier hypothesis") . Both DF, ADF and Phillips-Perron type corrected t-tests are computed (for details see Perron, 1989, for an application to EMS real exchange rates, see Delhausse and Urbain, 1991). These statistics are respectively denoted in Table 5.3 by τ, τ^* and Z. The 5% critical value depends on the location of the breaking point relative to the complete sample, i.e. on the so- called breaking fraction (Perron, 1989) T/T_B where T is the whole sample and T_B the date of the potential structural break. This breaking fraction approximately equals 0.40 so that the 5% the critical values are respectively −3.72, −3.94 and −4.22 for the Crash hypothesis (denoted by C in the Table), the Changing Growth (denoted by CG) hypothesis and both nested within a general model (C,CG). The results are reported in Table 5.3.

For imports as well as for the income variable we are now able to reject the null of a unit root at the conventional 5% level (denoted by the subscript "*" in the table). Does this mean that these series have to be considered stationary around a segmented trend function, so that cointegrated models become useless? If we could eventually answer yes in a purely univariate analysis, we must keep in mind that these univariate statistics are mainly used for descriptive purposes, and that Perron's procedure is relying on an important interpretation of the "structural break". In particular, it implies

		Additive outlier			Innovation Outlier
Variable	Case	τ	τ^*	Z	τ
			Belgium		
m	C	-3.753*	-2.431	-4.229*	-0.922
	CG	-5.378*	-2.845	-6.016*	-3.140
	C,CG	-7.406*	-2.843	-6.044*	-2.655
y	C	-3.698	- 2.316	-3.834*	-1.944
	CG	-4.179*	-2.261	-3.678	-2.035
	C,CG	-4.455*	-2.224	-4.038	-2.089
pm	C	-1.419	-2.227	-1.150	-2.237
	CG	-0.375	-1.864	-0.231	-1.451
	C,CG	-1.334	-2.042	-1.234	-1.868
pd	C	-2.003	-1.951	-1.953	-2.422
	CG	- 0.554	-1.806	-0.302	-1.631
	C,CG	-2.025	-1.919	-1.997	-2.239

Table 5.3: Testing for breaking trend: the 1973 oil shock

that the shock is purely deterministic and exogenous so that, in the case of the rejection of the unit root hypothesis, it can be viewed as **the only existing permanent shock** to the series. Notice that the assumption that 1973 is the only break in the series is doubtful given the within-sample evolution of recursive ADF tests, see Figures 5.8-5.11. These show high variability so that, depending on the selected sample, both rejection and non-rejection of the unit root hypothesis can occur. Consequently, we could also interpret 1973 as a shock, among others, whose permanent effect is more important than other permanent shocks which affect the series in the case of unit root processes.[3] Therefore, in the subsequent analysis we shall not

[3]We may also view this problem as an illustration of the crucial importance of the conditioning information set which underlies any empirical model. On a similar point, we note the study by Spanos (1990b) which investigates the conditions under which a unit root in some model does, or does not, carry over to models with a different conditioning information set. Similarly, as pointed out in Franses (1991b), it is possible that unit roots at the univariate level do not carry over to the multivariate level. In the procedure described in Perron (1989), the conditioning information set underlying the regressions

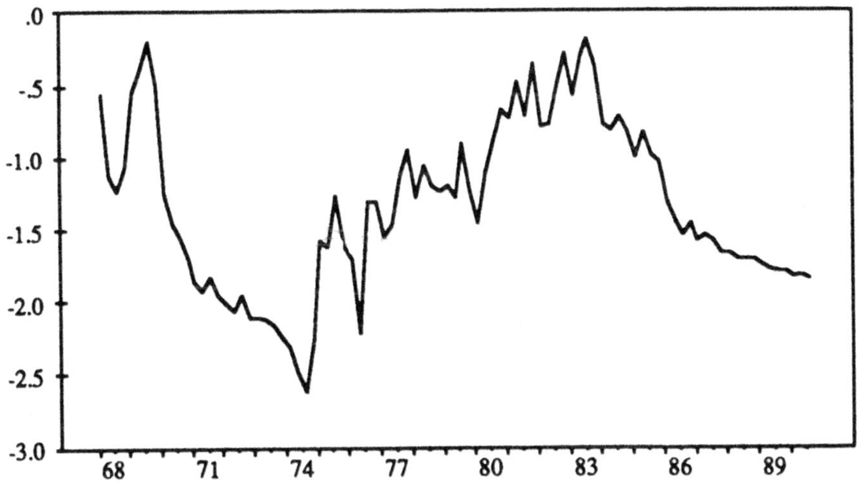

Figure 5.8: Sequential ADF tests: import volume

take this evidence as being against the $I(1)$ character of aggregate imports and income, but rather as about the importance of the effect of the 1973 oil shock. The more sensible strategy to follow, according to our empirical experience, is to consider the univariate results as rough data descriptions while the nonstationarity properties of the variables should more usefully be investigated in a cointegration framework (for a similar idea, see Lubrano, 1991) based on an extended conditioning information set.

is composed of past values of the series and some deterministic factors. Among these, a dummy for the oil shock has an important effect. One could however argue that an extended conditioning information set would not need this dummy if it includes a variable directly reflecting the important increase in oil prices in its historical behaviour. This is obviously the case with the aggregate import price index(Δpm) which, by construction, includes oil prices and shows an important upward peak in 1973. Surprisingly, although more directly affected by the 1973 oil-shock, the import price index does not provide any evidence against the unit root hypothesis.

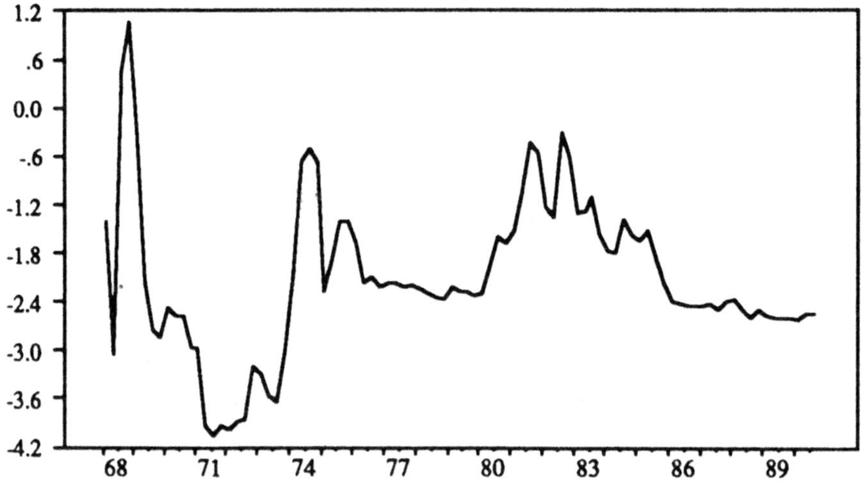

Figure 5.9: Sequential ADF tests: real income

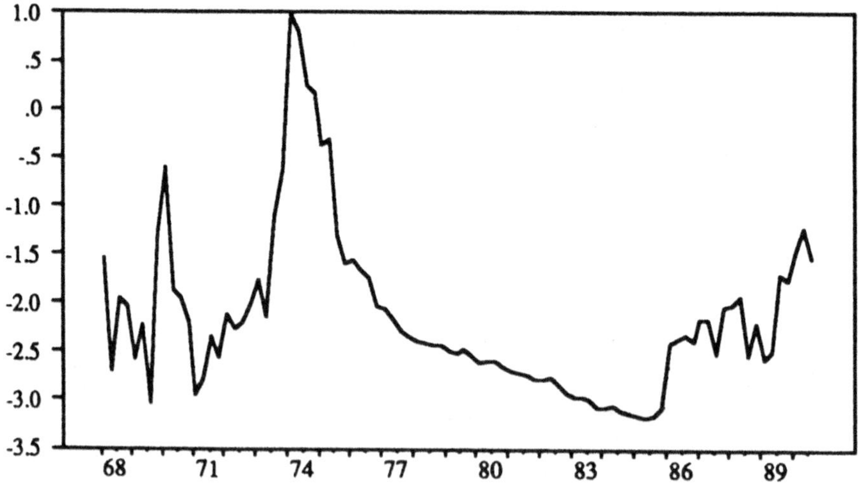

Figure 5.10: Sequential ADF tests: domestic prices

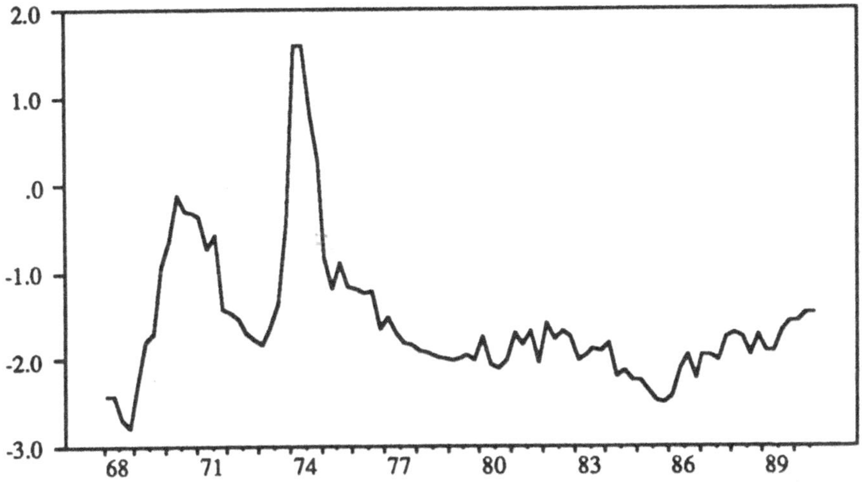

Figure 5.11: Sequential ADF tests: import prices

5.2.2 Multivariate cointegration and simultaneous equation models

The next step is to build a multivariate model enabling us to investigate the cointegration properties of the data and parametrise them in a more interpretable and economically meaningful model. Since we found no evidence in favour of $I(2)$ processes, we may safely conduct the analysis under the assumption that our selected variables are well represented by $I(1)$ processes with potential deterministic factors.

As has been pointed out in Chapter 2, we have several alternatives for the elaboration of cointegrated models. In a first step, we consider a **closed model** for aggregate imports and follow the approach proposed in Hendry and Mizon (1990). From a practical point of view, their recent approach is a generalisation of the Hendry methodology to systems of equations with non-stationary variables. The starting parametric model we use is a VAR model which, if congruent with the data evidence, is considered (see Spanos,1988, 1990a) as a characterization of the so-called Haavelmo distribution, i.e. a statistical model for the joint dependent variables denoted by the (4×1) vector x_t, with $x_t' = (m_t, y_t, pm_t, pd_t)$.[4]

[4]We limit ourselves here to the set of variables which are usually selected for aggregate

We should be aware here that our framework, although apparently general and unrestricted, is relying on the implicit assumption that the class of VAR models is general enough to approximate the unknown DGP. As pointed out in Palm (1990), marginalisation w.r.t. (un)observables may result in moving average terms, in which case some of our results would no longer be valid.

Testing for cointegration in VAR

The first step of the empirical analysis is conducted in the space of $I(1)$ variables where the integration and potential cointegration properties are taken into account. To this end, we use the maximum likelihood estimation procedure developed in Johansen (1988), Johansen and Juselius (1990) based on a complete VAR model in vector autoregressive error correction form:

$$\Delta x_t = \sum_{i=1}^{n-1} \Gamma_i \Delta x_{t-i} + \Gamma_n x_{t-n} + \mu + \varepsilon_t \tag{5.4}$$

where ε_t denotes a k-dimensional normal variate with mean zero and non-singular, p.d.s. covariance matrix Σ and μ is vector of constant terms. Given the univariate evidence concerning the seasonality in the data, the model is fitted with three centered deterministic seasonal quarterly dummies. Note also, from the figures in Appendix C, that the data series display clear linear trend components, so that we allow a non-zero deterministic drift vector μ to be present in the model.

The choice of the maximum lag length used in the specification of the vector error correction model (5.4) can affect the determination of the dimension of the cointegrating space, i.e. the rank of Γ_n, as pointed out in Boswijk and Franses (1992), Park and Ogaki (1991) and Reimers (1991). In particular, Boswijk and Franses (1992) and Reimers (1991) find that overfitting implies a loss of power while underfitting leads to potential spurious cointegration.

The problem that we encounter with the modelling of trade flows, and which appears as extremely specific to this type of data, is the long delayed

import studies since this is our main interest. Extending the information set to further variables increases the size of the model to estimate and moreover sometimes gives rise to hardly interpretable cointegration relationships.

effects of prices on imports. These may last up to two or three years (see the results reported in Goldstein and Khan, 1985; Urbain, 1990). As a matter of fact, fitting a 8th or higher order VAR implies too many degrees of freedom lost and very likely overfitting of the model as a whole. On the other hand, using lag order selection criteria such as Schwarz or AIC result in low order VAR models, usually of order two or three. If we simply use the VAR model for the determination and estimation of cointegrating relationships, these criteria are found in Boswijk and Franses (1992) and Reimers (1991) to perform relatively well in terms of size and power of cointegration tests. Our problem is however here different for we are trying to develop a first stage model that has to be general enough to capture the long run relations, but also the short run dynamic which will afterwards be more parsimoniously reparametrised. Consequently, we deliberately use VAR's of relatively high order to be certain that the short run dynamic is well captured.

Following the device of Boswijk and Franses (1992), we apply Johansen's procedure for different numbers of candidate models, allowing the lag length to vary from 3 to 7. The sensitivity of the outcome to the specification of the lag length is reported in Table 5.4 where the subscripts "*" and "**" denote significant at respectively 5% and 10% level. The last columns report the outcome of specific hypotheses we have tested on the cointegrating space. H_1 is the null hypothesis of long run price homogeneity (i.e. symmetry of the price effects) while H_2 is the strong null hypothesis of zero long run price effects (i.e. both zero domestic and import prices effects).

lag order (k)	Number of CI-vector (r) Trace	Max. Eig.	H_1 $\chi^2(r)$	H_2 $\chi^2(2.r)$
3	1^*	1^{**}	1.925	1.774
4	1^*	1^*	2.779	5.069
5	1^*	1^*	1.751	4.818
6	1^{**}	0	0.079	3.807
7	1^{**}	0	0.349	3.353

Table 5.4: Lag length selection and Johansen's cointegration tests

Although the number of cointegrating vectors differs according to the lag length selected, it is reassuring that the estimates of their components (when normalised on the diagonal) are very similar in almost all cases. It is also worth noting that the outcome of the parametric hypotheses we tested, on the cointegrating space, are the same irrespective of the lag length of the VAR. We based our final choice of the lag length on two criteria: (i) absence of serial correlation in the residuals (using Box-Pierce tests for 16th order serial correlation) and (ii) the significance of the parameter estimates of the short run coefficients (using F-tests for the nullity of the corresponding column in the VAR). This resulted in the choice of a lag length of 5 periods. Detailed results of the application of Johansen's multivariate cointegration approach are reported in Table 5.5. The 5% critical values (Johansen and Juselius, 1990) are reported in the tables of Appendix B.

Both the trace and the maximum eigenvalue tests detect at most one cointegrating relationship given by the first column of the β matrix which is normalised on its diagonal. The components of the cointegrating vector have the theoretical expected sign. Note that from the α matrix, the cointegrating relationship seems to enter essentially the first equation (that of Δm_t) giving some indications about the potential weak exogeneity of the remaining variables in the system, at least w.r.t. the long run parameters. A LR test for the null hypothesis of long run price homogeneity has also been conducted as shown in Table 5.6

Since the null hypothesis of long run price homogeneity (H_1) cannot be rejected, and this whatever the lag length, we present the estimated component of the restricted cointegrating vector under this hypothesis. The prices have relatively low coefficients with the "wrong" sign. The more restrictive hypothesis (H_2) that prices do not have significant effects in the long run is also not rejected. This last result is consistent with some of the single dynamic equation results reported in Urbain (1990, 1992a) where the long domestic price effect was found to be close to zero and where Johansen's approach did not reject long run price homogeneity.[5] This should not be considered as too surprising. Belgium is known to be extremely open and dependent on the foreign markets. About 65% of its import are composed of equipment goods and intermediate products not available on the national

[5] We have to note in that paper, that the analysis was based on annual data covering the period 1953-1985 and the hypothesis that prices do not enter the cointegrating relations was not formally tested, as was however the long run price homogeneity.

		k=5		
		Eigenvalues		
	0.2509	0.0993	0.0581	0.0233
		Eigenvectors		
m	-26.206	-6.470	13.885	2.897
y	30.939	-1.238	-10.691	4.638
pm	-5.172	-11.998	24.667	3.139
pd	13.389	27.481	- 40.186	- 9.540
		LR tests		
Trace Test		H_0	Max. Eigenvalue	
2.4141		$r \leq 3$	2.4141	
8.5205		$r \leq 2$	6.1065	
19.1876		$r \leq 1$	10.6671	
48.6580*		$r = 0$	29.4705*	
Normalised Cointegrating vectors: β matrix				
m	1.000	5.226	0.563	-0.304
y	- 1.181	1.000	-0.433	-0.491
pm	0.197	9.692	1.000	-0.329
pd	-0.511	-22.199	1.629	1.000
		Loading factors: α matrix		
m	- 0.229	0.010	-0.079	0.023
y	0.073	-0.003	-0.027	0.043
pm	0.098	0.001	-0.099	-0.018
pd	0.074	0.003	-0.011	-0.007

Table 5.5: Multivariate cointegration analysis

	H_1: price homogeneity LR test $(\chi^2(1)) = 2.663$	
	Restricted cointegrating vector	Restricted loading vector
m	1.000	-0.190
y	-1.386	0.075
pm	-0.179	0.077
pd	0.179	0.042
	H_2: zero long run price effects LR test $(\chi^2(2)) = 5.069$	
	Restricted cointegrating vector	Restricted loading vector
m	1.000	-0.226
y	-1.418	0.083
pm	0.000	0.045
pd	0.000	0.027

Table 5.6: Hypothesis tests on cointegrating matrix

markets, so that these imported goods are complementary products entering the production process and not substitutes for nationally produced goods. Prices will therefore certainly have a small effect in the long run on the Belgium external trade, or at least on aggregate imports (see the discussion in Italianer and d'Alcantara, 1986; Urbain, 1990).

The time series evolution of the restricted cointegrating vector is reported in Figure 5.12

In Hendry and Mizon's (1990) framework, the VAR model used in Johansen's approach plays the role of the unrestricted reduced form (hereafter labelled UVAR) model against which subsequent reductions and/or reparametrisations will be tested if the UVAR is a congruent *constant* parameter representation of the data. Similar to results reported in the VAR-based studies of Hendry and Mizon (1990) or Clements and Mizon (1991), the normality assumption is violated for the equation determining domestic prices. The equations for the other variables satisfy the normality assumption according to the outcome of Jarque and Bera's Normality test, distributed as a $\chi^2(2)$ under the null hypothesis of normality of the residuals. We conjecture

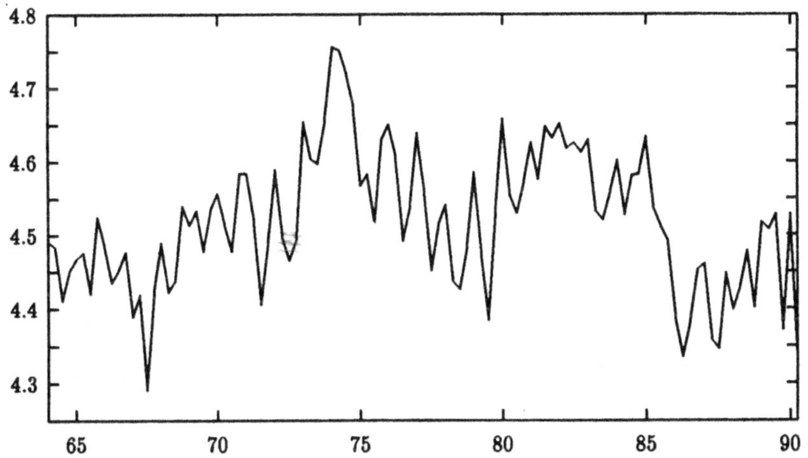

Figure 5.12: Restricted cointegrating vector

with these authors that non- normality should not alter Johansen's approach too much, insofar it is due to a limited number of outlier values.[6] As shown in the figures of the time series used in this study, the first differences of the price variables display important values at the end 1973 or early 1974, end 1979 as well as around the period 1985-1986. These do correspond respectively to the first and the second oil-shocks and to the important dollar rise of the mid-eighties. A within sample analysis of each equation of the UVAR, using multivariate recursive least squares and sequential Chow tests, clearly shows instability and outlier values around these periods. In order to remove these effects from our data set, following Hendry and Mizon (1990), Clements and Mizon (1991), we therefore used three dummies denoted by $D74$, $D79$ and $D86$. One must understand that the use of these dummies is due to the specific form of the model that we are maintaining. A closed VAR model does marginalise with respect to all the information in the variables (and their past values) not included, i.e. w.r.t. the variables which are probably the source of these parameter non-constancies and outlier values such as oil price, US exchange rates, ... However, introducing these variables induces an important increase in the size of the model[7] at the expense of less

[6]See Park and Ogaki (1991) for a different point of view.

[7]Preliminary results using oil price and US dollar exchange rate against the DTS led to less nonconstancy and nonnormality at the expense of an important loss of economic

interpretable results. As these dummies will enter each equation of the VAR, and hopefully only selectively in the final SEM, the task of encompassing the VAR model will be even more difficult. Notice that once these dummies are introduced, there remains almost no non-normality[8] (see Table 5.7) and sequential Chow tests revealed no significant within sample instability as shown in the Figures 5.12-14.

	UVAR			UVAR with dummies		
Var.	σ	BJ	BP	σ	BJ	BP
m	0.0346	2.537	12.3538	0.0329	0.397	17.2920
y	0.0288	5.712	8.3459	0.0264	0.140	12.9081
pm	0.0249	0.714	2.8722	0.0179	0.088	15.4136
pd	0.0089	38.836	4.9805	0.0078	3.642	11.2163

	PVAR			SEM		
Var.	σ	BJ	BP	σ	BJ	BP
m	0.0323	0.8891	16.7421	0.0324	0.866	16.0590
y	0.0315	0.6300	22.1040	0.0317	0.583	21.182
pm	0.0225	1.434	5.705	0.0228	0.960	3.749
pd	0.0069	0.575	9.424	0.0082	0.554	11.056

Test for the validity of overidentifying restrictions
$\chi^2(41) = 52.015$

Table 5.7: Statistics on the error processes

The unrestricted model is then reparametrised in the space of stationary $I(0)$ variables, resulting in a parsimonious VAR model, denoted by PVAR. The latter is specified by using first differences of the variables, dropping the variables which became redundant in the $I(0)$ space and dropping those for which appropriate F-test for their significance in each of the four equations

meaning of the statistical results so that this extension has not been further pursued.

[8]In Table 5.7, BJ is Jarque and Bera's normality test, distributed $\chi^2(2)$ under the null, BP is Box-Pierce test for 12th order serial correlation

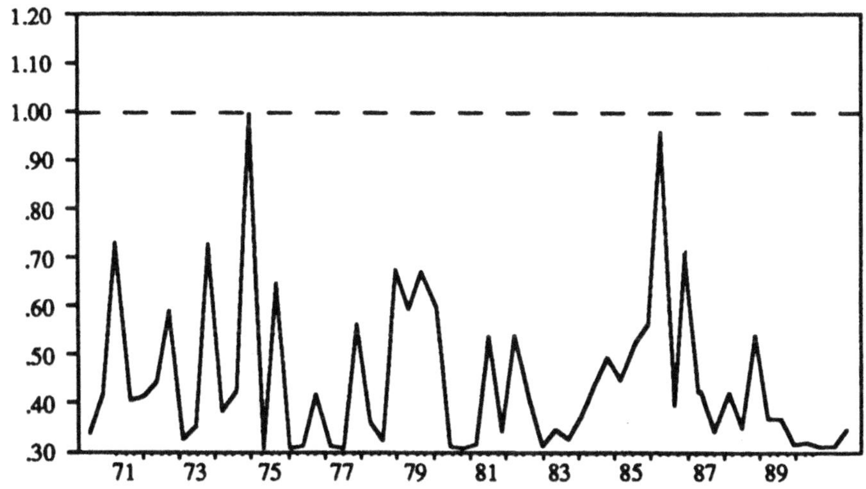

Figure 5.13: 1-step Chow tests (PVAR): import volume equation

is not significant. The PVAR is then checked for its constancy, normality and absence of serial correlation. If these properties are satisfied, the PVAR constitutes the general model against which the encompassing properties, i.e. the overidentifying restrictions, of the final SEM are tested. The statistics of the error processes of the UVAR, the UVAR with dummies, the PVAR and the selected SEM are displayed in Tables 5.7. The final SEM (estimated by FIML), presented in Table 5.8, has been derived by first adding as an identity, an empirically defined error correction term, denoted by ECM. For notational convenience, two further identities are added: $\Delta(m - y)$ and $\Delta\Delta pd$. This results in a SEM with seven endogenous among which three are identities. Note that the ECM term was defined accordingly to the results obtained in Johansen's approach. Following a general to simple modelling strategy we have been able to derive a parsimonious SEM which encompassed the corresponding PVAR, as shown at the bottom of Table 5.7 which reports encompassing tests against the corresponding PVAR. The latter statistic is simply a test for the validity of the overidentifying restrictions imposed by SEM.

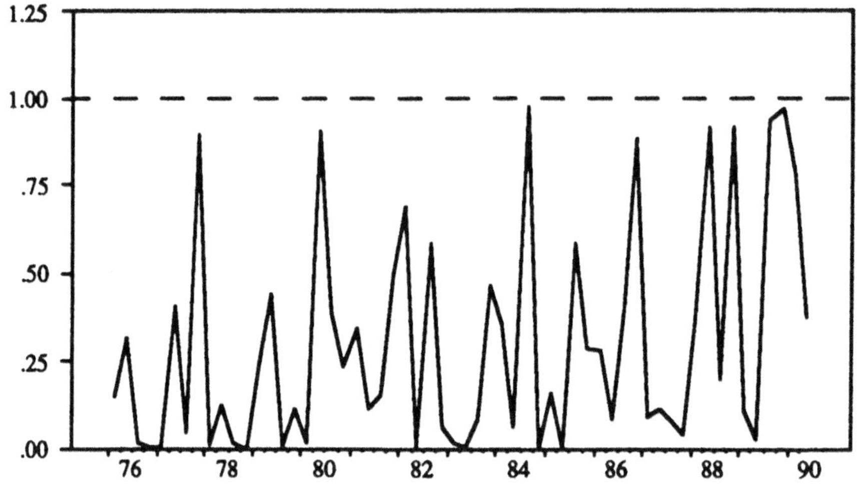

Figure 5.14: 1-step Chow tests (PVAR): real income equation

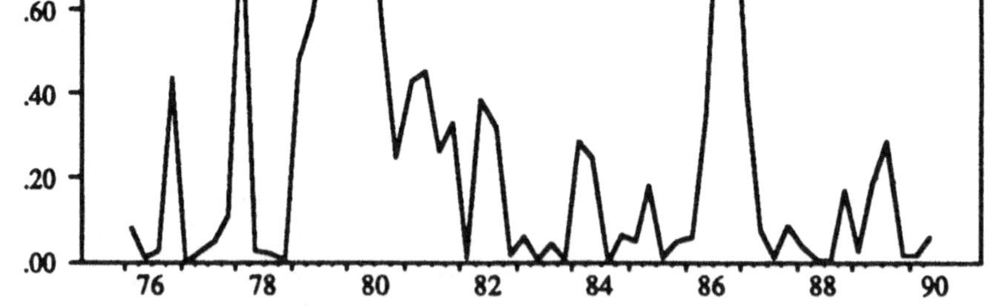

Figure 5.15: 1-step Chow tests (PVAR): domestic prices equation

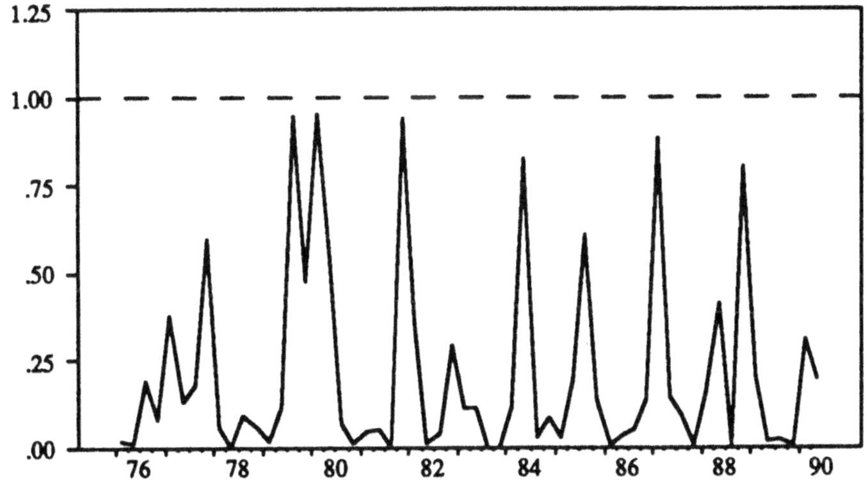

Figure 5.16: 1-step Chow tests (PVAR): import prices equation

If we focus on the first equation of the SEM, it corresponds to what is expected for import equations. In particular, as discussed earlier, domestic prices appear to have a more immediate and important effect than import prices whose effect may last up to five quarters. The use of a relative price thus appears clearly as inadequate if a well specified short run dynamic model for aggregate imports is to be derived. Note that for the prices and income equations, the presence of dummies is necessary in order to encompass the PVAR, giving some indication that our information set is insufficient to model correctly the behaviour of these variables.

A further interesting outcome of the SEM is its contrast with the unrestricted cointegrated VAR. In particular, the error correction term, which we have empirically defined and added to the systems as an identity, appears as significant in the equations for the domestic prices and income. This contrasts with the unrestricted loading factors estimated by the Johansen procedure and is certainly to be related to the important number of overidentifying restrictions which the final SEM implies. Our final SEM is a relatively parsimonious model, and has the nice property that it encompasses a densely parametrised VAR. It appears however sensitive to the use of dummies, so that the robustness of this modelling strategy might be questioned. Nevertheless, given the information set retained, and the empirical starting point, such a model has the advantage of capturing most of the dynamic present in the data and producing meaningful estimates of the short and long run price and income effects.

Var	Coeff	Std Err	Var	Coeff	Std Err
Equation for Δm_t			Equation for Δy_t		
$\Delta(m-y)_{t-1}$	-0.329	0.079	$D86$	0.039	0.223
$\Delta(m-y)_{t-2}$	-0.287	0.077	$D74$	-0.074	0.033
Δy_{t-3}	0.449	0.125	$D79$	-0.075	0.037
Δpd_t	1.310	0.495	Δpm_{t-2}	-0.431	0.130
Δpd_{t-3}	1.209	0.416	Δy_{t-3}	-0.181	0.094
Δpm_{t-3}	-0.417	0.184	ECM_{t-1}	0.131	0.041
Δpm_{t-5}	-0.452	0.146			
Δy_t	0.550	0.219			
ECM_{t-1}	-0.189	0.090			
Equation for Δpm_t			Equation for Δpd_t		
$D74$	0.098	0.023	$D86$	-0.015	0.006
Δpm_{t-1}	0.297	0.086	$D74$	0.128	0.011
$\Delta\Delta pd_{t-1}$	0.404	0.168	$\Delta\Delta pd_{t-2}$	0.219	0.058
Δpm_{t-2}	0.189	0.085	Δpm_t	0.399	0.073
			Δy_t	-0.109	0.058
			$\Delta(m-y)_{t-3}$	-0.041	0.014
			ECM_{t-1}	0.030	0.113
Definition of ECM term: $ECM_t = \Delta m_t - 1.404\Delta y_t + ECM_{1t-1}$					

Table 5.8: FIML estimation of the final SEM

5.2.3 Partial modelling of aggregate imports: a structural ECM

In the previous section we based our analysis on a complete system where the error correction term was added as an identity. Dummies were also required in order to have a well behaved global model with constant parameters and white noise errors. Since our interest lies first of all in modelling aggregate imports and not in deriving empirical models for wholesale prices or income, a conditional approach could equally well be suitable if the remaining variables in the system are weakly exogenous for both the short and long run parameters of the import model. In particular, economic theory has not a lot to tell about theoretical relationships between income, domestic and import prices, so that the remaining equations of the systems are mainly empirical models whose economic content is difficult to assess.

Following the approach presented in the Chapter 2 of this work, we now try to model aggregate imports using a partial-system framework where some of the variables are considered as weakly exogenous, so that their generating processes are not modelled. We consider that we have two endogenous variables which are import volume and import prices. Income and domestic prices are in a first stage considered as exogenous for the parameters of interest. As pointed out in the Chapter 2, this requires the use of identification restrictions in order to derive a well defined *structural* ECM with more than one (potential) endogenous variable.

In order to identify the long run import demand equation we normalise the first long run relationship such that imports has a unit coefficients and we also impose long run price homogeneity (denoted by R_1) as is usually done in theoretical presentations of import demand. For the price equation, it is less obvious, except for the normalisation, what restriction should be imposed on the cointegrating vectors. Remember that in this framework we assume that only one cointegrating vector is associated to each endogenous variable (see Chapter 2, Boswijk, 1991a, 1992a,b). Although theoretical models of import price setting usually involve further variables such as exchange rates, cost of production .. (see inter alia Kim, 1990; Yang, 1991; Marquez, 1990) some authors model import prices by restricting their information set to the variables that we are using (see Marquez and McNeilly, 1988), while studies of aggregate imports using rational expectation model prices simply by specifying univariate autoregressive models or VARMA models for income and prices (see Gagnon, 1988; Husted and Kollintzas, 1984; 1987). The fact that we are working in a total aggregate framework makes things more difficult for we are aggregating over both the commodities and the trade partners of Belgium. When bilateral data are studied, import prices are easier to model using for example the export price of the trading partner (in foreign currency), the bilateral exchange rate (see Marquez, 1990). In any case, aggregate import demand functions remain important in practice as they are often used in national macro-models which usually do not disaggregate their foreign sector.

In order to identify the price equation, we impose the restriction that income should not enter the cointegrating vector. Although this restriction does not have a profound theoretical justification, it can be argued that the exporters are fixing their prices according to volume of goods demanded as well as according to the prices that are fixed at the national level. This restriction is denoted by R_2.

We consider the following structural error correction model

$$\Gamma \Delta y_t = \kappa + \Pi_0 \Delta z_t + \lambda \gamma' x_{t-1} + \sum_{i=1}^{p-1} (\Phi_i \Delta y_{t-i} + \Pi_i \Delta z_{t-i}) + \eta_t \tag{5.5}$$

where Γ is 2×2 nonsingular matrix with unit diagonal elements, y_t is here a 2×1 vector of endogenous variables whose components are m_t and pm_t while z_t a 2×1 vector of weakly exogenous variables y_t and pd_t.

In order to identify the two *separate* long run relationships we have imposed restrictions of the type

$$R_i \gamma_i = e_i \quad i = 1, 2$$

where R_i and e_i are matrices with a priori known coefficients corresponding to the restrictions just mentionned, e.g. R_1, R_2 where with R_1 we impose price homogeneity to identify the long run demand and R_2 serves to identify the price equation. We also assume that the matrix of structural error correction coefficients (λ) is a diagonal matrix. In that case, each equation of this two-equation system contains the jointly dependent variables but only one error correction term, so that each endogenous variable is associated with only one long run relationship. We implicitly assume that we have at most two cointegrating relationships among our 4-dimensional set of variables. In order to test for cointegration in this framework, we started the analysis by fitting general ECMs where the maximum lag length was fixed at five periods in order to ensure the models to have residuals with the desired statistical properties. We used both instrumental variable (IV) and OLS estimation procedures, the latter being only valid if there is only one cointegrating vector and one endogenous variable. In the case of IV estimation, for equation i, the instruments selected are the predetermined variables in the model as well as $R_i x_{t-1}$ which, by definition, are valid instruments for equation i since they are absent from the equation i due to the identification restriction but correlated to the other endogenous due to the error correction term in the other equation. The presence of cointegration is tested using a Wald test for the significance of the error correction term. Note that a constant and a trend term are added to the model in order the ensure the asymptotic similarity of Boswijk's (1991a) Wald test for the significance of the error correction term. The results are reported in Table 5.9.

Dep.Var.	Method	Id. Rest.	$k = 5$
m	OLS	-	27.632
	IV	R_1	19.934
pm	IV	R_2	7.009

Table 5.9: Wald tests for Cointegration

The 5% critical values are tabulated in Boswijk (1989, 1991a) and approximately equal 17.20 in the case of two exogenous variables and 19.82 for three exogenous. Both with OLS and IV, we reject the null hypothesis of no cointegration in the import demand equation for we find a significant cointegrating vector. It must be emphasized that, within this framework, what we are testing is not only the null hypothesis of no-cointegration but also implicitly the validity of the identifying restriction. This points out an interesting characteristic of the approach advocated by Boswijk (1991a), which is that it allows to check directly *for the existence long run theoretically justified relationships*. In the case of Belgium, it appears that long run homogeneity cannot be rejected. In accordance with the evidence provided by the Johansen procedure, we do not reject the null of no-cointegration in the case of the import price equation since the error correction term does not appear to be significantly different from zero.

Var.	IV	OLS_{res}	OLS_{ures}
λ_1	-0.219(0.09)	-0.218(0.09)	-0.191(0.14)
m_{t1}	-1	-1	-1
y_{t-1}	1.324(0.26)	1.301(0.27)	1.402(0.28)
pr_{t-1}	0.079(0.21)	0.017(0.22)	-
pm_{t1}	-	-	-0.226(0.23)
pd_{t-1}	-	-	0.371(0.29)
Cst	4.876(0.43)	4.869(.44)	5.572(0.44)
Trend	0.0006(0.003)	0.0004(0.002)	0.00011(0.003)
η_1	2.60	1.69	2.42
η_2	1.88	1.57	2.04
η_3	1.19	1.10	1.44
η_4	1.03	0.89	0.97

Table 5.10: Cointegrating vector estimates

In Table 5.10, we present the estimated cointegrating relation both when IV or OLS is applied, under the restriction of long run price homogeneity and without any restriction in the OLS case. *pr* denoted relative prices, i.e. $pm - pd$. Remark that the reported standard errors, calculated using Bardsen's (1989) method, are only valid if weak exogeneity of y_t, pm_t and pd_t is verified. The bottom rows of the Table presents statistics on the error processes of the general models. $\eta_i, i = 1, 2, 3, 4$ are respectively Jarque and Bera's (1980) normality test, a LM test for fifth order serial correlation in F-form, an F-test for 4th order ARCH effects and White's (1980) functional form misspecification test.

The point estimates of the income coefficient are relatively independent of the estimation method used. It is worth noting that both the estimated error correction coefficients and estimated income effects are close to those obtained with annual data (see Urbain, 1990). In a similar way to what was observed in Johansen's approach, prices have "the wrong" sign when they are restricted to have symmetric effects in the long run. Without restrictions, they have the correct expected signs but are not significant, at least according to the reported standard errors. These estimates are based on unrestricted error correction models.

Although this is a purely finite sample point, we may expect more precise (efficient) estimates (especially the error correction coefficient) if the general model is parsimoniously reparametrised. Since we found no evidence in favor of a second endogenous variables, we conduct the analysis using OLS and use a set of tests statistics in order to be ensured that each restriction on the model, i.e. each reduction, is data acceptable. This resulted in a parsimonious ECM which imposes 12 restrictions on the general model. A global $F(12,34)$-test for their validity gives 1.03, well below the 1% critical value, allowing us to consider the reductions as reliable. The coefficient estimates, along with their heteroscedastic consistent standard errors, are reported in Table 5.11. η_i are test statistics on the properties of the error processes. η_1 is the usual Chow test for predictive failure, distributed as an $F(12,72)$ under the null, η_2 is Jarque and Bera's (1980) normality test, η_3 is the LM test for fifth order serial correlation distributed as an $F(5,68)$ under the null, η_4 is the LM test for 4th order ARCH effects distributed as an $F(4,69)$ under the null; η_5 is White's (1980) functional form misspecification test and η_6 is the second order RESET test, distributed as an $F(2,71)$ under the null.

Nevertheless, the validity of the computed standard errors hinges on the weak exogeneity assumption of the remaining variables in the system. As we consider both the short and long run dynamic response parameters of interest, weak exogeneity requires:

1. the absence of the cointegrating vector in the remaining equations,

2. the usual orthogonality condition.

From the discussion in the previous chapter, we therefore use a simple variable addition test for checking both hypotheses jointly. We fitted three unrestricted reduced ECMs for $\Delta y_t, \Delta pm_t, \Delta pd_t$ to which we added as an artificial regressor the fitted residuals from the structural ECM, while the error correction term is that estimated from the structural ECM. We only report their respective coefficients and standard errors along with some statistics on the error processes, as the other estimated coefficients are of no interest for our current purpose. No significant misspecification seems to affect these marginal models, both in terms of predictive failure and the (mean) innovation character of their residuals. Note however the observed non-normality of the residual from the domestic price equation.

The row *F-weak* is an F-test for the joint nullity of the coefficients of the added fitted residuals and the error correction term, i.e. an ***F-test for the null hypothesis of weak exogeneity*** of the corresponding variable for both the short and long run parameters of the import demand equation. It is distributed as an $F(2,71)$ under the null hypothesis, with a 5% critical value approximately equal to 3.10. As shown in the Table 5.11, we cannot reject the hypothesis that income, import prices and domestic prices are all weakly exogenous variables for our parameters of interest.

Var	Δm_t	Δpm_t	Δy_t	Δpd_t
Cst.	0.907(0.37)			
S1	-0.068(0.02)			
S2	-0.024(0.01)			
S3	-0.169(0.02)			
Trend	0.0003(0.001)			
Δy_t	0.445(0.131)			
$\Delta(m-y)_{t-1}$	-0.331(0.08)			
$\Delta(m-y)_{t-2}$	0.295(0.07)			
Δy_{t-3}	0.438(0.10)			
Δpd_t	1.287(0.34)			
Δpd_{t-3}	1.207(0.49)			
Δpm_{t-3}	-0.440(0.22)			
Δpm_{t-5}	0.469(0.14)			
ECM	-0.183(0.08)	0.059(0.06)	0.108(0.07)	0.034(0.03)
$\hat{v}_t$		0.024(0.08)	- 0.051(0.10)	0.004(0.03)
R^2	0.87	0.66	0.72	0.66
σ	0.035	0.029	0.034	0.011
DW	2.06	2.14	2.08	1.98
η_1	0.92	1.24	1.61	2.45
η_2	3.12	0.90	2.63	25.77
η_3	1.00	0.86	1.07	0.47
η_4	1.22	0.66	0.36	0.29
η_5	1.01	0.78	0.74	0.69
η_6	0.55	1.16	0.09	2.321
$F-weak$	-	0.55	0.96	0.85
ECM		$m_t - 1.149y_t -$ (0.24)	$0.039(pm-pd)_{t-1}$ (0.19)	

Table 5.11: Structural ECM and weak exogeneity

We may compare these results with the outcome of LR test statistics for weak exogeneity that one can conduct within Johansen's framework. We first estimated the cointegrating space under linear homogeneous restrictions on the cointegrating matrix (where we impose price homogeneity) and on the loading matrix. The maximised likelihood function under these hypotheses is then compared to the unrestricted maximised likelihood using standard LR tests, asymptotically χ^2 distributed. The results are summarised in Table 5.12 where both individual weak exogeneity as well as joint weak exogeneity are tested, but now with respect to the long run parameters only.

H_0	pm_t	y_t	pd_t
Weak Exo. ($\chi^2(2)$)	3.093	5.829	4.754
Joint Weak Exo. ($\chi^2(4)$)		6.870	

Table 5.12: Weak exogeneity tests

None of the results enable us to reject the null of weak exogeneity of r.h.s. variables of the import demand equation for the long run parameters. Reliable inference can thus be drawn using the computed standard errors(see inter alia Johansen, 1992a; Boswijk, 1992b). In Table 5.11, the standard error of the price effect is relatively high compared to the point estimate, so that long run price effects might well be negligible. A formal test of this hypothesis is not rejected at any convenient significance level. Therefore the final selected parametrisation takes the form of the single equation structural error correction model reported in Table 5.13.

The reported statistics do not show any sign of misspecification so that this ECM can be considered as a data acceptable representation of the underlying DGP.

Variable	Coefficient	Standard Error
$Constant$	0.909	0.35
S_1	-0.066	0.03
S_2	0.023	0.02
S_3	- 0.164	0.03
$Trend$	0.0004	0.0004
Δy_t	0.442	0.12
$\Delta(m-y)_{t-1}$	-0.332	0.09
$\Delta(m-y)_{t-3}$	-0.296	0.03
Δpd_t	1.283	0.33
Δpd_{t-3}	1.200	0.46
Δpm_{t-3}	-0.438	0.16
Δpm_{t-5}	-0.469	0.14
ECM_{t-1}	-0.243	0.11
R^2	0.87	
DW	2.02	
σ	0.035	
η_1	0.97	
η_2	3.28	
η_3	0.93	
η_4	1.32	
η_5	1.14	
η_6	0.54	
$ECM_t =$	$m_t -$	$1.111\ y_t$ (0.23)

Table 5.13: ECM for aggregate imports

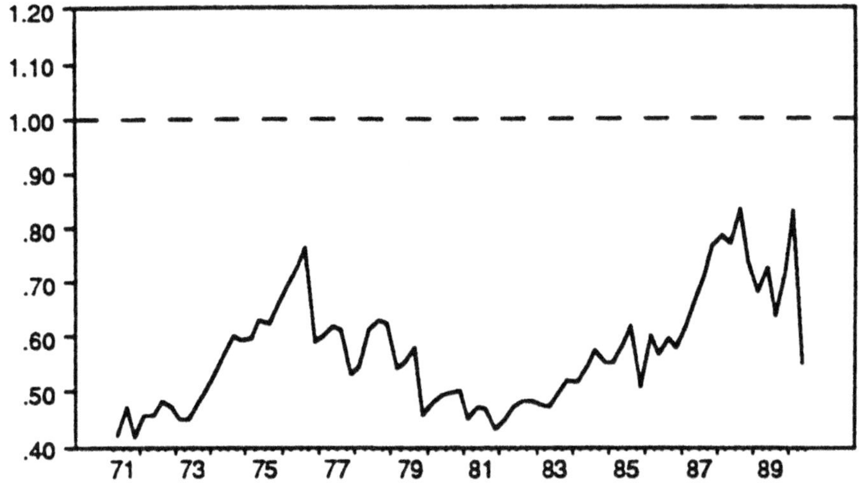

Figure 5.17: N-step ahead Chow tests: structural ECM

A final check of this specification is conducted using a recursive estimation procedure in order to test for the within sample stability of the model. As pointed out for example by Engle and Hendry (1990), this is one of the simplest ways to check for the super exogeneity character of the conditioning variables, if we do not want to use parametric test statistics for super exogeneity. In fact, if the (condtional) model appears to have stable coefficients through time, although it is known that the conditioning variables have been affected by important changes over the sample period, then we may not reject the hypothesis that the model is structurally invariant to changes affecting the generating processes of the conditioning variables. In other words, we may not reject the hypothesis that the conditioning variables are super exogenous for the parameters of interest, so that valid policy simulations can be performed using the model (for a recent overview of these concepts in relation to cointegrated models, see Ericsson, 1992). We therefore recursively estimated our final selected error correction model. Sequential Chow tests revealed no significant structural instability as shown in figures 5.17, 5.18, 5.19. Remark that each calculated statistics have been scaled by their 5% critical value, so that a crossing of the horizontal unit line implies a rejection of the underlying null hypothesis of coefficients stability.

A similar recursive estimation of the marginal ECMs for the condition-

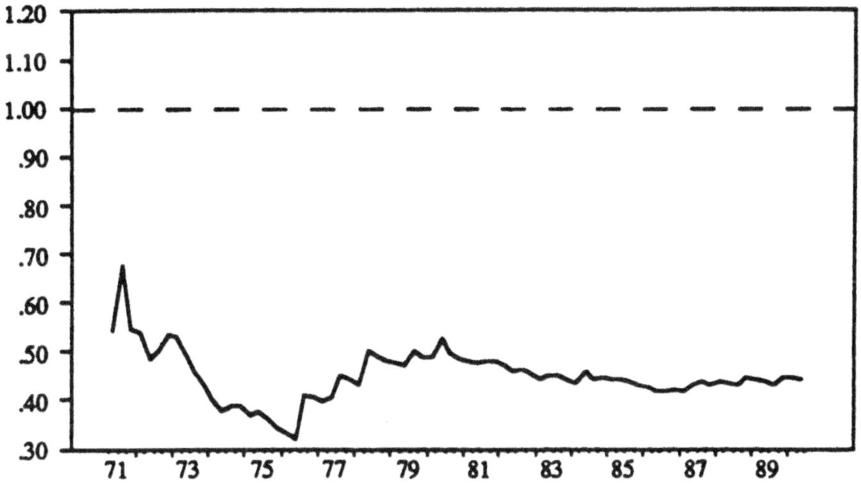

Figure 5.18: N-step backward Chow tests: structural ECM

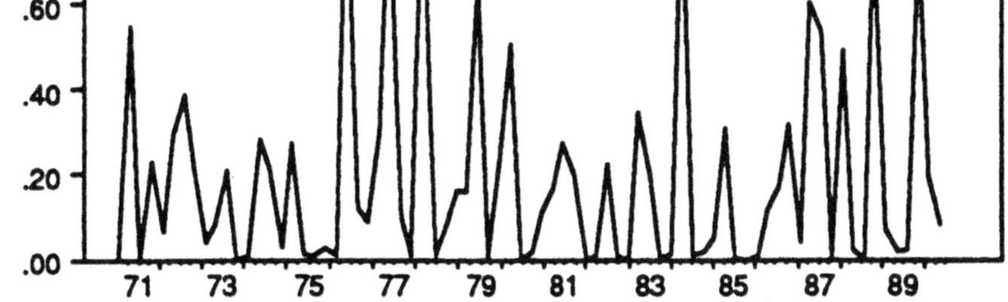

Figure 5.19: 1-step Chow tests: structural ECM

ing variables points out subtantial instability in all of our marginal models. However, as we do not detect any significant within sample instability affecting the ECM for aggregate imports, we may safely consider that the data does not provide any evidence against the specification retained.

5.3 Conclusions

A detailed interpretation of the results we have obtained goes far outside the scope of this work, but it is interesting to note that the results which we obtain in this structural ECM framework are somewhat different to those using the Johansen approach.[9] In particular, the long run income effect is lower, almost 1.1 in comparison to 1.4, while the prices do not appear to have a significant effect in the long run. This last result, corroborating results reported in Urbain (1990), implies that the Belgium economy is extremely dependent on the foreign markets. A possible explanation is, as already mentioned, the important share of equipment goods, durable and intermediate products in total aggregate imports.

This also has important implications in terms of the effectiveness of, for example, exchange rate policies and, at least according to the evidence provided in this study, casts serious doubts on the effectiveness of devaluation policies. With respect to the use of a relative price ratio, the empirical evidence provides a further argument against the use of such a ratio in the short run as shown in the point estimates. Domestic prices, as expected, have a more important effect, in terms of size, which apparently occurs more immediately. In particular, the short run effect of domestic price changes is almost three times greater than that of import prices changes and so contributes to the important dependent character of the Belgium economy. Short run distortions imply an important differentiation of the prices effects and should clearly be accounted for in any attempt to conduct economic policies affecting, directly or indirectly, the trade balance.

---

[9] The zero long run price effect remains nevertheless a puzzling result. While this result is "robust" to various estimation methods, it seems however dependent on the choice of the price terms used in the analysis (see Urbain, 1992d).

Chapter 6

Conclusions

In the recent years, the study of cointegrated times series and the use error correction models have become extremely popular in the applied as well as in the theoretical econometric literature. In this field of study, the choice between a full-system based approach and a partial approach using sub-systems is not clear-cut, and is even open to debate. If some exogeneity conditions are verified, lower dimension systems are obviously, from a practical point of view, easier to handle and from a theoretical point of view, enjoy nice asymptotic properties. The empirical researcher is often confronted with the choice of running the risk of losing efficiency or losing empirical tractability. As exogeneity is essential for being able to conduct valid inference in conditional cointegrated systems, we have investigated the conditions of weak exogeneity within a particularly useful representation of cointegrated systems which is the popular error correction model.

Our main findings can be summarised as follows.

1. System based approaches, while theoretically and asymptotically appealing, have their own practical problems: problems of the size of the system when an important number of variables is introduced, problems of interpretation, lack of economic theory to build "complete" specifications, ... This makes the use of partial (conditional) systems attractive for empirical research. The validity of these however relies on the exogeneity status of the variables on which we condition. As weak exogeneity is dependent on the choice of the parameters of interest,

the framework used in the empirical modelling of cointegrated systems must be made precise. In particular, we found that weak exogeneity of r.h.s. variables in a conditional ECM necessitates the absence of the cointegrating vector in the generating models of the conditioning variables, a condition which is satisfied when the endogenous does not Granger cause the r.h.s. variables.

2. In many cases, the applied econometrician prefers to introduce more "structure" on his/her model in order to get economically meaningful coefficients. For this purpose, ECMs in structural form provide an appealing framework, allowing the researcher to introduce (theoretically motivated) identification restrictions on the long run relationships. Also, short run dynamic response parameters may be of great help in desrcibing economic phenomena and fit particularly well this "structural framework". In this case, weak exogeneity requires more than the absence of the cointegrating vector in the generating models. As is usual in structural models, a traditional orthogonality condition has also to be satisfied.

3. Orthogonality tests have often been used in order to test for weak exogeneity or for general misspecification. In cointegrated systems, they seem to behave poorly in finite samples, with an additional complication that can arise from nonstandard limiting distribution. Further, the null hypothesis underlying orthogonality tests, is not sufficient for weak exogeneity unless the cointegrating vector is known to be absent from the generating model of the potentially exogenous variables. Formal tests for the latter hypothesis have recently been proposed, and it is straightforward to extend such tests by considering *jointly* the orthogonality condition. This results in a simple variable addition test for weak exogeneity. The implication in terms of representation of incomplete simultaneous equation models is also more clear-cut as only one of these representations allows to derive such a test for weak exogeneity in a relatively easy way.

We have illustrated the theoretical analysis in a study of Belgium's aggregate imports over the period 1964-1990. Using quarterly data, we followed both a complete system-based approach as well as a partial approach where we have allowed for the introduction of identifying information on the cointegrating vectors. Weak exogeneity of the conditioning variables was tested using the procedure described above.

Our major empirical finding is that a single equation ECM for Belgian aggregate imports emerges as empirically well behaved (parameter constancy, no dynamic misspecification, valid conditioning variables, ...). Since weak exogeneity is not rejected for both the long run and short dynamic response parameters, it allows reliable inference on the crucial problem of price and income effects in international trade. In comparison with the system based analysis, the structural ECM seems more suited for this problem as we are first of all interested in modelling the import (and eventually import prices) behaviour. The closed system approach also shows excessive sensitivity to the use of dummies, a point which may cast doubts on the robustness of this modelling strategy for data such as used in aggregate import demand models. The derived price and income effects are in accordance with a previous study based on annual data using a different sample period. In particular, the price homogeneity assumption, often imposed in empirical models for aggregate imports, is clearly inadequate to describe the short run response of imports to changes in import and domestic prices. More interestingly, we found that the open and dependent character of the Belgium economy on the foreign markets, results in the non rejection of the hypothesis that prices do not matter in the long run. As this last result appears independent to the estimation method, it might have important implications for the effectiveness of economic policies aimed to affect directly or indirectly the trade balance of Belgium.

Given our empirical experience within this framework, we want to conclude this work by describing some arguments in favor of a progressive empirical modelling strategy where full systems and partial systems emerge as complementary in the modelling process, and not substitutes. Conditional models have clearly a number of practical advantages which make them extremely appealing for empirical purposes. As we are often interested in a reduced number of relationships which implicitly take the form of conditional relationships, these conditional models reduce the danger of misspecification which is inherent in complete systems. Economic theory is often of little guide for complete "structural specifications". Nonetheless, a complete system approach should not be considered as an alternative to partial system approach, but well as a convenient tool for having a general and reliable statistical description of the data, often more informative for the derivation of econometric models than simple univariate statistics such as unit root tests. For example, starting the analysis of integrated time series with the Johansen's procedure allows the empirical researcher to have a nice full account of the potential problems that can occur if he wants to restrict his attention to a subset of the variables. The Johansen procedure allows him to discover the number of long run relationships (cointegrating vectors), while the unrestricted (or restricted) estimates of the factor loadings provide a rapid indication on the presence of the cointegrating vectors in the various equations of the system. Finally, a first check on the validity of restrictions on the long run relationships is also easily obtained in this framework. When more structure is imposed, a partial approach has the advantage of minimising the risk of misspecification of the model for the conditioning variables. Since weak exogeneity is testable in this framework, one may conduct a more detailed specification and still check for the validity of the resulting estimates and standard errors in due course.

We do certainly not pretend to have solved the issue or to have addressed the problem completely. We hope, at least, to have pointed out that when the conditions for weak exogeneity are clearly stated and their validity checked, structural (conditional) ECMs appear appealing dynamic presentations of economic phenomena. Further work is naturally desirable. In particular, additional empirical studies of trade flows using this framework should be conducted[1] in order to really asses to what extend this modelling framework solves (or avoids) the usual problems that have been encountered in traditional econometric analysis of trade flows. An important step is also to derive a theoretical justification for the use of ECMs for aggregate imports, in which case the data mining critic, which is always applicable to derived specifications, would no more be valid.

[1] A similar analysis has been conducted for five other countries. It leads to different estimates and parametrisations but nevertheless seems to provide strong support for conditional models. In particular, the sensitivity of Hendry and Mizon's (1990) framework to shift dummies is observed in almost all cases, leading to the use of up to 11 dummies in the case of France.

Bibliography

[1] Aldrich, J. (1989), "Autonomy", **Oxford Economic Papers**, 41, 15-34.

[2] Alogoskoufis, G. and R. Smith (1991), "On Error Correction Models: Specification, Interpretation, Estimation", **Journal of Economic Surveys**, 5, 97-128.

[3] Anderson, T.W. (1991), "Trygve Haavelmo and Simultaneous Equation Models", **Scandinavian Journal of Statistics**, 18, 1-19.

[4] Anderson, G.J. and G.E. Mizon (1990), "What Can Statistics Contribute to the Analysis of Economic Structural Change ?" in: P. Hackl (ed.), **Statistical Analysis and Forecasting of Economic Structural Change**, Springer Verlag, Berlin.

[5] Asseery, A., and D.A. Peel (1991), "Estimates of a Traditional Aggregate Import Demand Model for Five Countries", **Economic Letters**, 35, 435-439.

[6] Banerjee, A., J.J. Dolado , D.F. Hendry and G. Smith (1986), "Explor ing Equilibrium Relationships in Econometrics through Static Models: some Monte Carlo Evidence", **Oxford Bulletin of Economics and Statistics**, 48, 253-277.

[7] Banerjee, A., J.W. Galbraith and J.J. Dolado (1990), "Dynamic Specification and Linear Transformations of the Autoregressive-Distributed Lag Model", **Oxford Bulletin of Economics and Statistics**, 52, 95-104.

[8] Banerjee, A. and D.F. Hendry (1992), "Testing for Integration and Cointegration: An Overview", **Oxford Bulletin of Economics and Statistics**, 54, 225-256.

[9] Bärdsen, G. (1989), "Estimation of Long Run Coefficients in Error Correction Models", **Oxford Bulletin of Economics and Statistics**, 51, 345-350.

[10] Barndorff-Nielsen, O. (1978), **Information and Exponential Families in Statistical Theory**, John Wiley, Chichester.

[11] Bauwens, L. and H.K. van Dijk (1990), "Bayesian Limited Information Analysis", in: J.J. Gabszewicz, J.F. Richard and L.A. Wolsey (eds.), Revisited", **Economic Decision - Making: Games, Econometrics and Optimisation**, Amsterdam, North Holland, 385–424.

[12] Billinglsey, P. (1968), **Convergence of Probability Measures**, John Wiley and Sons, New York.

[13] Boswijk, H.P. (1989), "Estimation and Testing for Cointegration with Trended Variables : a Comparison of a Static and a Dynamic Regression Procedure", Report AE 12/89, Institute of Actuarial Science and Econometrics, University of Amsterdam.

[14] Boswijk, H.P. (1990), "On the Scope of Conditional Dynamic Modelling of Cointegrated Variables", **Tinbergen Institute Research Bulletin**, 2, 97-111.

[15] Boswijk, H.P. (1991a), "Testing for Cointegration in Structural Error Correction Models" Report AE 7/91, Institute of Actuarial Science and Econometrics, University of Amsterdam.

[16] Boswijk, H.P. (1991b), "The LM test for Weak Exogeneity in Error Correction Models" Report AE 13/91, Institute of Actuarial Science and Econometrics, University of Amsterdam.

[17] Boswijk, H.P. (1992a), "Efficient Inference on Cointegration Parameters in Structural Error Correction Models", Report 10/92, Institute of Actuarial Science and Econometrics, University of Amsterdam.

[18] Boswijk, H.P. (1992b), **Cointegration, Identification and Exogeneity**, Ph.D. Thesis/Tinbergen Institute, Amsterdam.

[19] Boswijk, H.P. and F.H. Franses (1992), "Dynamic Specification and Cointegration", **Oxford Bulletin of Economics and Statistics**, 54, 369-381.

[20] Boswijk, H.P. and J.-P. Urbain (1991), "The Behaviour of Orthogonality Tests in Cointegrated Models", mimeo, University of Liège.

[21] Box, G.E.P. and G.M. Jenkins (1970), **Time Series Analysis, Forecasting and Control**, Holden-Day, San Fransisco.

[22] Campbell, J.Y. and R.J. Shiller (1987), "Cointegration and Tests of Present Value Models", **Journal of Political Economy**, 95, 1062-1088.

[23] Campbell, J.Y. and R.J. Shiller (1988), "Interpreting Cointegrated Models", **Journal of Economic Dynamics and Control**, 12, 505-522.

[24] Campbell, J. and P. Perron (1991), "Pitfalls and Opportunities : What Macroeconomists should know About Unit Roots", NBER Technical Working Paper 100, Cambridge.

[25] Campbell, J. and P. Perron (1992), "Racines Unitaires en Macroéconomie: le Cas Multidimensionnel", **Annales d'Economie et de Statistique**, 27, 1-50.

[26] Capuccio, N. and R. Orsi (1990), "Testing Exogeneity in Overindentified Models", CORE Discussion Paper 9042.

[27] Cheng, H.S. (1959), "Statistical Estimates of Elasticities and Propensities in International Trade: a Survey", **IMF Staff Papers**, 7, 107-158.

[28] Chow, G. (1960), "Tests of Equality Between Sets of Coefficients in Two Linear Regression Models", **Econometrica**, 28, 591-605.

[29] Chow, G. (1990), "The Multiplier-Accelerator Model in the Light of Cointegration", Research Memorandum 357, Princeton University.

[30] Clarida, R.H. (1991), "Co-integration, Aggregate Consumption and the Demand for Imports: a Structural Econometric Investigation", NBER Working Paper 3812, Cambridge.

[31] Clements, M.P. (1990), "The Mathematical Structure of Models that Exhibit Cointegration: a survey of recent approaches", Applied Economics Discussion Paper 85, Nuffield College, Oxford.

[32] Clements, M.P. and G.E. Mizon (1991), "Empirical Analysis of Macroeconomic Time Series : VAR and Structural Models", **European Economic Review**, 35, 887-932 (with comments).

[33] Cuthbertson, K., S.G. Hall and M.P. Taylor (1992), **Applied Econometric Techniques**, Phillip Allen, New York.

[34] Davidson, J. (1986), "Error Correction Systems", mimeo, London School of Economics.

[35] Davidson, J. (1991), "The Cointegration Properties of VAR models", **Journal of Time Series Analysis**, 12, 41-61.

[36] Davidson, J. and S. Hall (1991), "Cointegration in Recursive Models", **Economic Journal**, 101, 239-251.

[37] Davidson, J.E.H., D.F. Hendry, F. Sbra and S. Yeo (1978), "Econometric Modelling of the Aggregate Time-Series Relationship Between Consumers Expenditure and Income in the United Kingdom", **Economic Journal**, 88, 661-692.

[38] Delhausse, B. and J.-P. Urbain (1990), "Purchasing Power Parity, Productivity and Terms of Trade : a Multivariate Cointegration Analysis", CREDEL Research paper 9002 (Revised), University of Liège.

[39] Delhausse, B. and J.-P. Urbain (1991), "Times Series Behaviour of Exchange Rates, the Unit Root Hypothesis and the Creation of the EMS", mimeo, University of Liège.

[40] Dhrymes, P.J. (1981), **Distributed Lags, Problems of Estimation and Formulation**, North Holland, Amsterdam (second edition).

[41] Dickey , D. A. and W.A. Fuller (1979), "Distribution of the Estimators for Autoregressive Time Series with a Unit Root", **Journal of the American Statistical Association**, 74, 427-431.

[42] Dickey , D.A. and W.A. Fuller (1981), "Likelihood Ratio Test Statistics for Autoregressive Time Series with a Unit Root", **Econometrica**, 49, 1057-1072.

[43] Diebold, F.X. and M. Nerlove (1990), "Unit Roots in Economic Time Series : A Selective Survey", in: Fomby, T.B. and G.F. Rhodes (eds.), **Advances in Econometrics : Cointegration, Spurious Regression and Unit Roots**, JAI Press.

[44] Dinter, U. (1990), "The Autoregressive Distributed Lag System with Cointegrated Non-Stationary I(1)-variables", mimeo, Fern Universität Hagen.

[45] Dolado, J.J. (1991), "Asymptotic Distribution Theory for Econometric Estimation with Integrated processes: A guide", Documento de Trabajo 9103, Bank of Spain.

[46] Dolado, J.J. (1992), "A Note on Weak Exogeneity in VAR Cointegrated Models", **Economics Letters**, 38, 139-143.

[47] Dolado J.J., T. Jenkinson and S. Sosvilla-Rivero (1991), "Cointegration and Unit Roots", **Journal of Economic Surveys**, 4, 249-273.

[48] Dolado, J.J., N.R. Ericsson and J. Kremers (1989), "Inference in Conditional Dynamic Models with Integrated variables", mimeo, paper presented at ESEM 1989, Munich.

[49] Domowitz, J. and M. Hakkio (1990), "Interpreting an Error Correction Model: Partial Adjustment, Forward Looking Behaviour and Dynamic International Demand for Money", **Journal of Applied Econometrics**, 5, 29-46.

[50] Engle, R.F. (1984), "Wald, Likelihood Ratio and Lagrange Multiplier Tests in in: Z. Griliches and M.D. Intriligator, (eds.), Econometrics", **Handbook of Econometrics** Vol. 2, North Holland, Amsterdam.

[51] Engle, R.F., Hendry, D.F. and Richard, J.F. (1983), "Exogeneity", **Econometrica**, 51, 277-304.

[52] Engle, R.F. and Granger, C.W.J. (1987), "Co-integration and Error Correction: Representation, Estimating and Testing", **Econometrica**, 55, 251-276.

[53] Engle, R.F. and S. Yoo (1987), "Forecasting and Testing in Cointegrated Systems", **Journal of Econometrics**, 35, 143-159.

[54] Engle, R.F. and D.F. Hendry (1990), "Testing Super Exogeneity and Invariance", mimeo, Nuffield College, Oxford University.

[55] Engle, R.F. and B.S. Yoo (1989), "Cointegrated Economic Times Series: A Survey with New Results", in: R.F. Engle and C.W.J. Granger (eds.), **Modelling Long-run Economic Relationships: Readings in Cointegration**, Oxford University Press, Oxford.

[56] Epstein, R.J. (1988), **A History of Econometrics**, North Holland, Amsterdam.

[57] Ericsson, N.R. (1992), "Cointegration, Exogeneity and Policy Analysis", **Journal of Policy Modelling**, 14, 251-280.

[58] Ericsson, N.R., J. Campos and H.A. Tran (1991), "PC-GIVE and David Hendry's Econometric Methodology", mimeo, **Revista de Econometria**, forthcoming.

[59] Escribano, A. (1987), "Co-integration, Time Co-Trends and Error Correction Systems: An Alternative Approach", CORE Discussion paper 8715.

[60] Evans, G.B.A. and N.E. Savin (1984), "Testing for Unit Roots: II", **Econometrica**, 52, 1241- 1268.

[61] Florens, J.-P. and M. Mouchart (1985), "Conditioning in Dynamic Models", **Journal of Time Series Analysis**, 6, 15-34.

[62] Franses, P.H. (1991a), "Seasonality, Nonstationarity and the Forecasting of Monthly Time Series", **International Journal of Forecasting**, 7, 199-208.

[63] Franses, P.H. (1991b), **Model selection and Seasonality in Time Series**, Ph.D. thesis, Erasmus University Rotterdam.

[64] Fuller, W.A. (1976), **Introduction to Statistical Times Series Analysis**, John Wiley, New York.

[65] Gagnon, J. (1988), "Adjustment Costs and International Trade Dynamics", **Journal of International Economics**, 26, 327-344.

[66] Geweke, J. (1984), "Inference and Causality in Economic Time Series" in: Z. Griliches and M.D. Intriligator (eds.), **Handbook of Econometrics**, Vol. 2, North Holland, Amsterdam.

[67] Ghysels, E. and P. Perron (1990), "The Effect of Seasonal Adjustment Filters on Tests for a Unit Root", mimeo, University of Montreal.

[68] Gilbert, C.L. (1986), "Professor Hendry's Econometric Methodology", **Oxford Bulletin of Economics and Statistics**, 48, 283-307.

[69] Goldstein, M. and M. Khan (1985), "Income and Price Elasticities in Foreign Trade", in: R. Jones and P.Kenen (eds.), **Handbook of International Economics**, Vol.2, North Holland, Amsterdam.

[70] Gonzalo, J. (1989), "Comparison of Five Alternative Methods of Estimating Cointegrating Vectors", mimeo, University of California at San Diego.

[71] Gourieroux, C. (1990), "Quelques Développements Récents en Séries Temporelles, **Journal de la Société Statistique de Paris**, 1, 8-22.

[72] Gourieroux, C. and A. Monfort (1989), **Statistique et Modèles Econométriques**, Vol. 1&2, Paris, Economica.

[73] Gourieroux, C. and A. Monfort. (1990), **Séries Temporelles et Modèles Dynamiques**, Paris, Economica.

[74] Granger, C.W.J. (1981), "Some Properties of Time Series Data and their use in Econometric Model Specification", **Journal of Econometrics**, 16, 121-130.

[75] Granger, C.W.J. (1990), ed. **Modelling Economic Series**, Oxford University Press, Oxford.

[76] Granger, C.W.J. and P. Newbold (1974), "Spurious Regression in Econometrics", **Journal of Econometrics**, 26, 1045-1066.

[77] Granger, C.W.J. and A. Weiss (1983), "Time Series Analysis of Error-Correction Models", in: S. Karlin, T. Amemiya and L.A. Goodman (eds.), **Studies in Econometrics, Time series and Multivariate statistics**, Academic Press, New York.

[78] Granger, C.W.J. and P. Newbold (1986), **Forecasting Economic Time Series**, Academic Press, San Diego (second edition).

[79] Gregoire, S. and G. Laroque (1991), "Multivariate Integrated Time Series: a General Error Correction Representation with Associated Estimation and Test Procedure", mimeo, INSEE Paris.

[80] Gregory, A.W. (1991), "Testing for Cointegration in Linear Quadratic Models", Discussion Paper, Queens University.

[81] Harvey, A.C. (1981), **Time Series Models**, Phillip Allen, Oxford.

[82] Harvey, A.C. (1989), **The Econometric Analysis of Time Series**, Phillips Allen, New York (2nd edition).

[83] Hausman, J.A. (1978), "Specification Tests in Econometrics", **Econometrica**, 46, 1251-1272.

[84] Hausman, J.A. (1984), "Specification and Estimation of Simultaneous Equation Models" in: Z. Griliches and M.D. Intriligator (eds.), **Handbook of Econometrics**, Vol. 1, North Holland, Amsterdam.

[85] Haynes, S. and J. Stone (1983), "Secular and Cyclical Response of U.S. Trade to Income: An Evaluation of Traditional Models", **Review of Economics and Statistics**, 65, 87-95.

[86] Haynes, S. and J. Stone (1985), "A Neglected Method of Separating Demand and Supply in Time Series Regressions", **Journal of Applied Econometrics**, 3, 238-244.

[87] Hendry, D.F. (1983), "Econometric Modelling : The Consumption Function in Retrospect", **Scottish Journal of Political Economy**, 30, 193-220.

[88] Hendry, D.F. (1986), "Econometric Modelling with Cointegrated Variables", **Oxford Bulletin of Economics and Statistics**, 48, 201-212.

[89] Hendry, D.F. (1987), "Econometric Methodology: a Personal Perspective" in Bewley, T. (ed.), **Advances in Econometrics - Forth World Congress**, Vol.1 , Cambridge University Press, London, 29-48.

[90] Hendry, D.F. (1988), "The Encompassing Implications of Feedback versus Feedforward Mechanisms in Econometrics", **Oxford Economic Papers**, 40, 132-149.

[91] Hendry, D.F. (1992), "On Interactions of Unit Roots and Exogeneity", mimeo, Nuffield College.

[92] Hendry, D.F. and T. von Urgern-Stenberg (1981), "Liquidity and Inflation Effects on Consumers Behaviour", in: A. Deaton (ed.), **Essays**

in the Theory and Measurement of Consumers' Behaviour, Cambridge University Press, Cambridge.

[93] Hendry, D.F. and J.F. Richard (1982), "On the Formulation of Empirical Models in Dynamic Econometrics", **Journal of Econometrics,** 20, 3-33.

[94] Hendry, D.F. and J.F. Richard (1983), "The Econometric Analysis of Economic Time Series" (with discussion), **International Statistical Review,** 51, 111-163.

[95] Hendry, D.F. and G.E. Mizon (1990), "Evaluating Dynamic Models by Encompassing the VAR" in: P.C.B. Phillips and V.B. Hall (eds.), **Models, Methods, and Applications of Econometrics, Essays in Honor of Rex Bergstrom,** Basil Blackwell, forthcoming.

[96] Hendry, D.F. and N. Ericsson (1991), "Assertion without Empirical Basis: an Econometric Appraisal of Monetary trends in the U.K... by M. Friedman and A. Schwartz", **American Economic Review,** 81, 8-38.

[97] Hendry, D.F., A.R. Pagan and J.D. Sargan (1984), "Dynamic Specification" in: Z. Grilliches and M.D. Intriligator (eds.), **Handbook of Econometrics,** Vol.2, North Holland, Amsterdam.

[98] Hendry, D.F., A.J. Neale and F. Sbra (1988), "Econometric Analysis of Small Linear Systems using PC-FIML", **Journal of Econometrics,** 38, 203-226.

[99] Holly, A. (1983), "Une présentation Unifiée des Tests d'Exogénéité dans les Modèles Simultanés", **Annales de l'INSEE,** 50, 3-24.

[100] Holly, A. (1985), "Testing for Exogeneity : a Survey", mimeo, Cahier de Recherches Economiques, Université de Lausanne.

[101] Horvarth, B. (1991), **Are Policy Variables Exogenous ?,** Lecture Notes in Economics and Mathematical Systems 364, Springer Verlag, Berlin.

[102] Hunter, J. (1990), "Cointegrating Exogeneity", **Economics Letters,** 34, 33-35.

[103] Hsiao, C. (1984), "Identification", in: Z. Grilliches and M.D. Intriligator (eds.), **Handbook of Econometrics**, Vol.1, North Holland, Amsterdam.

[104] Husted, S. and T. Kollintzas (1984), "Import Demand with Rational Expectations: Estimates for Bauxite, Cocoa, Coffee and Petroleum", **Review of Economics and Statistics**, 66, 608-618.

[105] Husted, S. and T. Kollintzas (1987), "Linear Rational Expectations Equilibrium Laws of Motion for Selected U.S. Raw Materials Imports", **International Economic Review**, 28, 651-670.

[106] Hylleberg, S. and G.E. Mizon (1989), "Cointegration and Error Correction Mechanisms", **Economic Journal**, 99 (Conference), 113-125.

[107] Hylleberg, S., R.F. Engle,C.W.J. Granger and B.S. Yoo (1990), "Seasonal Integration and Cointegration", **Journal of Econometrics**, 44, 215-238.

[108] Italianer, A. (1987), "Estimation et Simulation du Système des échanges Internationaux de Biens du Modèle QUEST", **Economie Européenne**, 63-131.

[109] Italianer, A. and d'Alcantara (1986), "Modelling Bilateral Sectoral Trade flows", in: P. Artus and O. Guvenen (eds.), **International Macroeconomic Modelling for Policy Decision**, Martinus Mijhof, Dordrecht.

[110] Jarque, C. and A. Bera (1980), "Efficient Tests for Normality, Homoskedasticity and Serial Independence of Regression Residuals", **Economics Letters**, 6, 255-259.

[111] Johansen, S. (1988), "Statistical Analysis of Cointegration Vectors", **Journal of Economic Dynamics and Control**, 12, 231-254.

[112] Johansen, S. (1990a), **Likelihood Based Inference on Cointegration: Theory and Applications**, Centro Interuniversitario di Econometrica, Bologna.

[113] Johansen, S. (1990b), "A Representation of Vector Autoregressive Processes integrated of order 2", Preprint 3, Institute of Mathematical Statistics, University of Copenhagen, **Econometric Theory**, forthcoming.

[114] Johansen, S. (1991a), "Estimation and Hypothesis Testing of Cointegration in Vectors Gaussian Autoregressive Models", **Econometrica**, 59, 1551-1580.

[115] Johansen, S. (1991b), "Estimating Systems of Trending Variables", mimeo, University of Copenhagen, **Econometric Reviews**, forthcoming.

[116] Johansen, S. (1992a), "Cointegration in Partial System and the Efficiency of Single Equation Analysis", **Journal of Econometrics**, 52, 389-402.

[117] Johansen, S. (1992b), "Determination of Cointegrating Rank in the Presence of a Linear Trend", **Oxford Bulletin of Economics and Statistics**, 54, 383-398.

[118] Johansen, S. and K. Juselius (1990), "Maximum likelihood Estimation and Inference on Cointegration - With Applications to the Demand for Money", **Oxford Bulletin of Economics and Statistics**, 52, 169-210.

[119] Johansen, S. and K. Juselius (1992a), "Testing Structural Hypotheses in a Multivariate Cointegration Analysis of the PPP and UIP for UK", **Journal of Econometrics**, 53, 211-244.

[120] Johansen, S. and K. Juselius (1992b), "Identification of the Long Run and Short Run Structure: an Application to the ISLM Model", mimeo, University of Copenhagen.

[121] Johnston, J.J. (1984), **Econometric Methods**, Mac Graw Hill, New York (third edition).

[122] Junz, H.B. and R.R. Rhomberg (1965), "Prices and Export Preformances of Industrial Countries: 1953-1963", **IMF Staff Papers**, 12, 224-269.

[123] Keil, M.W. and Richardson (1990), "A Comparison among Partial Adjustment, Rational Expectations and Error Correction Estimates of the Canadian Demand for Money", **Journal of Applied Econometrics**, 5, 273-291.

[124] Kholi, U. (1990), "Price and Quantity Elasticities in U.S. Foreign Trade", **Economic Letters**, 33, 277-281.

[125] Kim, Y. (1990), "Exchange Rate Pass-Through in the United States: A Varying Parameter Estimation of Exchange rate Pass-Through", **Journal of Business and Economic Statistics**, 8, 305-3156.

[126] Kim, K. and P. Schmidt (1990), "Some Evidence on the Accuracy of Phillips-Perron Tests using Alternative Estimates of Nuisance Parameters", **Economic Letters**, 34, 345-350.

[127] King, R.G., C. Plosser and S. Rebelo (1988), "Production, Growth and Business Cycles II", **Journal of Monetary Economics**, 21, 309-341.

[128] Kirchgässner, G. (1991), Comments on Clements and Mizon (1991), **European Economic Review**, 35, 918-922.

[129] Kiviet, J. (1986), "Model Selection Test Procedures in a Single Linear Equation of a Dynamic Simultaneous System and their Defect in Small Sample", **Journal of Econometrics**, 28, 327-362.

[130] Kloek, T. (1984), "Dynamic Adjusment when the Target is Nonstationary", **International Economic Review**, 25, 315-326.

[131] Kollintzas, T.E. and S. Husted (1984), "Distributed Lags and Intermediate Goods Imports", **Journal of Economic Dynamic and Control**, 8, 303-327.

[132] Koopmans, T. (1950), ed. **Statistical Inference in Dynamic Economic Models**, Cowles Commission Monograph 10, Wiley, New-York.

[133] Koopmans, T., H. Rubin and R.B. Leipnik (1950), "Measuring the Equation Systems of Dynamic Economics", in: T. Koopmans (ed.), **Statistical Inference in Dynamic Economic Models**, Cowles Commission Monograph 10, Wiley, New-York.

[134] Koopmans, T. and W. Hood (1953), eds. **Studies in Econometric Methods**, Cowles Commission Monograph 14, Yale University Press, New Haven.

[135] Krämer, W. (1985), "The Hausman Test with Trended Variables", **Economics Letters**, 14, 23-30.

[136] Kremers, J.J.M., N.R. Ericsson and J.J. Dolado (1992), "The Power of Cointegration Test", **Oxford Bulletin of Economics and Statistics**, 54, 325-248.

[137] Lahaye, J.M. and J.-P. Urbain (1992), "Structural Invariance and Super Exogeneity in Macroeconometric Model Building", **Cahiers Economiques de Bruxelles**, 134, 209-234.

[138] Leamer, E.E. and R. Stern (1971), **Quantitative International Economics**, Allyn and Baccon, Boston.

[139] Liu, T.C. (1960), "Underidentification, Structural Estimation and Forecasting", **Econometrica**, 28, 855-865.

[140] Lubrano, M., R. Pierce and J.F. Richard (1986), "Stability of a UK Demand for Money: a Bayesian Approach to Testing Exogeneity", **Review of Economic Studies**, 53, 603-637.

[141] Lubrano, M. (1991), "Testing for Unit Roots and Cointegration in a Bayesian Framework", DP9103, Université Catholique de Louvain-La-Neuve.

[142] Lucas, R. (1976), "Econometric Policy Evaluation : a Critique ", in: K. Brunner and A.H. Metzler (eds.), **The Phillips Curve and Labour Markets**, Carnegie-Rochester Conference Series on Public Policy, 2, (conference supplement), North Holland

[143] Lütkepohl, H. (1991), **Introduction to Multiple Time Series Analysis**, Spinger Verlag, Berlin.

[144] Magee, S. (1975), "Prices, Income and Foreign Trade", in: P. Kenen (ed.), **International Trade and Finance: Frontiers for Research**, Cambridge University Press, Cambridge.

[145] Malinvaud, E. (1981), "Econometrics Faced with the Needs of Macroeconomic Policy", **Econometrica**, 49, 1363-1371.

[146] Malinvaud, E. (1988), "Econometric Methodology at the Cowles Commission: rise and maturity", **Econometric Theory**, 4, 187-209.

[147] Marquez, J. (1990), "Bilateral Trade Elasticities", **Review of Economics and Statistics**, 72, 70-77.

[148] Marquez, J. (1991), "The Econometrics of Elasticities or the Elasticity of Econometrics: An Empirical Analysis of the Behaviour of U.S. Imports", mimeo, Board of Governors of the Federal Reserve System.

[149] Marquez, J. and C. McNeilly (1988), "Income and Price Elasticities for Exports of Developing Countries", **Review of Economics and Statistics**, 70, 306-314.

[150] Mayes, G. (1981), **Applications of Econometrics**, Prentice Hall, London.

[151] Mizon, G.E. (1984), "The Encompassing Approach in Econometrics" in: D.F. Hendry and K.F.Wallis (eds.), **Econometrics and Quantitative Economics**, Basil Blackwell, Oxford.

[152] Murray, T. and P.J. Ginman (1976), "An Empirical Examination of the Traditional Aggregate Import Demand Model", **Review of Economics and Statistics**, 58, 75-80.

[153] Monfort, A. and R. Rabemanajara (1990), "From a VAR model to a Structural Model : with an Application to the Wage Price Spiral", **Journal of Applied Econometrics**, 5, 203-227.

[154] Mosconi, R. and Giannini (1992), "Non Causality in Cointegrated Systems: Representation, Estimation and Testing", **Oxford Bulletin of Economics and Statistics**, 54, 399-417.

[155] Nakamura, A and M. Nakamura (1981), "On the Relationship among Several Specification Error Tests Presented in Durbin, Wu and Hausman", **Econometrica**, 49, 15831-588.

[156] Nelson, C.R. and C.I. Plosser (1982), "Trends and Random Walks in Macroeconomic Time Series : some Evidence and Implications", **Journal of Monetary Economics**, 10, 139-162.

[157] Newey, W. K. and K.D. West (1987), "A Simple Positive Definite, Heteroskedasticity and Autocorrelation Consistent Covariance Matrix" , **Econometrica**, 55, 703-708.

[158] Nickell, S. (1985), "Error Correction, Partial Adjustment and all that: an Expository Note", **Oxford Bulletin of Economics and Statistics**, 47, 119-129.

[159] Ogaki, M. and J.Y. Park (1990), "A Cointegrating Approach to Estimating Preference Parameters" mimeo, Cornell University.

[160] Oliviera, J. (1990), "Estimation des Flux du Commerce Extérieur: Nouvelles Approaches et Implications pour la Politique Economique", **Economie et Prévision**, 94-95, 197-216.

[161] Ouliaris, S. , Park, J.Y. and P.C.B. Phillips (1988), "Testing for a Unit Root in the Presence of a Maintained Trend", Working Paper 88-32, Cornell University.

[162] Orcutt, G. (1950), "Measurement of Price Elasticities in International Trade", **Review of Economics and Statistics**, 32, 117-132.

[163] Pagan, A. (1987), "Three Econometric Methodologies : A critical appraisal", **Journal of Economic Surveys**, 1, 3-24.

[164] Palm, F.C. (1983), "Structural Econometric Modelling and Time Series Analysis: an Integrated Approach", in: A. Zellner (ed.), **Applied Time Series Analysis of Economic Data**, Washington DC, US Department of Commerce, Bureau of the Census, 199-233.

[165] Palm, F.C. (1990), "Dynamic Econometric Modelling of Decisions under Uncertainty", in: F. van der Ploeg (ed.), **Advanced Lectures in Quantitative Economics**, Academic Press, London.

[166] Palm, F.C. and J.M. Sneek (1984), "Significance Tests and Spurious Correlation in Regression Models with Autocorrelated Errors", **Statistical Papers**, 25, 87-105.

[167] Palm, F.C. and G.A. Pfann (1990), "Interrelated Demand Rational Expectations Models for two typed of Labour", **Oxford Bulletin of Economics and Statistics**, 52, 45-68.

[168] Palm, F.C. and G.A. Pfann (1991), "Interrelation, Structural Change and Cointegration in a Model for Manufacturing Demand in the Netherlands", **Recherches Economiques de Louvain**, 57, 221-243.

[169] Park, J.Y. (1990), "Maximum Likelihood Estimation of Simultaneous Cointegration Models", Memo-1990-18, University of Aarhus.

[170] Park, J.Y. (1992), "Canonical Cointegrating Regressions", **Econometrica**, 69, 119-143..

[171] Park, J.Y. and P.C.B. Phillips (1988), "Statistical Inference in Regressions with Integrated Processes : Part I", **Econometric Theory**, 5, 468-497.

[172] Park, J.Y. and P.C.B. Phillips (1989), "Statistical Inference in Regressions with Integrated Processes : Part II", **Econometric Theory**, 6, 95-131.

[173] Park, J.Y. and M. Ogaki (1991), "Inference in Cointegrated Models using VAR Prewhitening to Estimate Short Run Dynamics", mimeo, Cornell University.

[174] Pesaran, H.M. and R. Smith (1990), "A Unified Approach to Estimation and Orthogonality Tests in Linear Single Equation Econometric Models", **Journal of Econometrics**, 44, 41-66.

[175] Perron, P. (1989), "The Great Crash, the Oil Price Schock and the Unit Root Hypothesis", **Econometrica**, 57, 1361-1402.

[176] Phillips, P.C.B. (1986), "Understanding Spurious Regressions in Econometrics", **Journal of Econometrics**, 33, 311-340.

[177] Phillips, P.C.B. (1987), "Time Series Regressions with a Unit Root", **Econometrica**, 55, 277-301.

[178] Phillips, P.C.B. (1988a), "Reflections on Econometric Methodology", Cowles Foundation Discussion Paper 893, Yale University.

[179] Phillips, P.C.B. (1988b), "Regression Theory for Near-Integrated Times Series", **Econometrica**, 56, 1021-1043.

[180] Phillips, P.C.B. (1991a), "Optimal Inference in Cointegrated systems", **Econometrica**, 59, 283-306.

[181] Phillips, P.C.B. (1991b), "To Criticise the Critics: An Objective Bayesian Analysis of Stochastic Trends" (with comments and reply), **Journal of Applied Econometrics**, 6, 333-364.

[182] Phillips, P.C.B. and S. Durlauf (1986), "Multiple Regression with Integrated Processes", **Review of Economic Studies**, 57, 99-125.

[183] Phillips, P.C.B. and P. Perron (1988), "Testing for a Unit Root in Time Series Regression", **Biometrika**, 75, 335-346.

[184] Phillips, P.C.B. and S. Ouliaris (1988), "Testing for Cointegration using principal component methods", **Journal of Economic Dynamic and Control**, 12, 205-230.

[185] Phillips, P.C.B. and B.E. Hansen (1990), "Statistical Inference in Instrumental Variables Regression with $I(1)$ Processes", **Review of Economic Studies**, 57, 99-125.

[186] Phillips, P.C.B. and S. Ouliaris (1990), "Asymptotic Properties of Residual Based Tests for Cointegration", **Econometrica**, 58, 165-193.

[187] Phillips, P.C.B. and M. Loretan (1991), "Estimating Long Run Equilibria", **Review of Economic Studies**, 58, 407-436.

[188] Phillips, P.C.B. and Y. Toda (1991), "Vector Autoregression and Causality", Cowles Foundation Discussion paper 977.

[189] Poirier, D. (1990), " A Conversation on Econometric Methodology" (D.F. Hendry, E.E. Leamer and D. Poirier)", **Econometric Theory**, 6, 171-261.

[190] Reimers, H.E. (1991), "Comparison of Multivariate Cointegration Tests", **Statistical Papers**, forthcoming.

[191] Richard, J.F. (1979), "Exogeneity, Inference and Prediction in So-Called Incomplete Dynamic Simultaneous Equation Models", CORE Discussion paper 7922.

[192] Richard, J.F. (1980), "Models with Several Regimes and Changes in Exogeneity", **Review of Economic Studies**, 77, 1-20.

[193] Richard, J.F. (1984), "Classical and Bayesian Inference in Incomplete Simultaneous Equation Models", in: D.F. Hendry and K.F. Wallis (eds.), **Econometrics and Quantitative Economics**, Basic Blackwell, Oxford.

[194] Saikkonen, P. (1991a), "Asymptotically Efficient Estimation of Cointegration Regressions", **Econometric Theory**, 7, 1-21.

[195] Saikkonen, P. (1991b), "Estimation of Cointegration Vectors in a Simultaneous Equation Framework", mimeo, paper presented at ESEM91 in Cambridge.

[196] Sargan, J.D. (1964), "Wages and prices in the United Kingdom : A study in econometric Methodology", in: P.E. Hart, G. Mills and J.K. Whitaker (eds.), **Econometric Analysis for National Economic Planning** , Butterworths, London.

[197] Sargent, T. (1981), "Interpreting Economic Times Series", **Journal of Political Economics**, 89, 213-248.

[198] Savin, N.E. (1984), "Multiple Hypothesis Testing", in: Z. Griliches and M.D. Intriligator (eds.), **Handbook of Econometrics**, Vol.2, North Holland, Amsterdam.

[199] Schwert, W. (-1989), "Tests for Unit Roots : A Monte Carlo Investigation", **Journal of Business and Economic Statistics**, 7, 147-159.

[200] Schotman, P. and H.K. van Dijk (1991), "A Bayesian Analysis of the Unit Root in Real Exchange rates", **Journal of Econometrics**, 49, 195-238.

[201] Sims, C. (1980), "Macroeconomics and Reality", **Econometrica**, 48, 1-48.

[202] Sims, C. (1991), Comments on Clements and Mizon, **European Economic Review**, 35, 922-932.

[203] Spanos, A. (1986), **Statistical Foundation of Econometric Modelling**, London, Cambridge University Press.

[204] Spanos, A. (1988), "Towards a Unifying Methodological Framework for Econometric Modelling", **Economic Notes**, 1, 1-28, reprinted as Chapter 15 in: C.W.J. Granger (1990).

[205] Spanos, A. (1990), "The Simultaneous Equation Model Revisited : Statistical Adequacy and Identification", **Journal of Econometrics**, 44, 87-105.

[206] Spanos, A. (1990b), "Unit Roots and Their Dependence On The Conditioning Information Set", in: Fomby, T.B. and G.F. Rhodes (eds.), **Advances in Econometrics : Cointegration, Spurious Regression and Unit Roots**, JAI Press, 271-293.

[207] Steel, M.J. (1987a), "Testing for Exogeneity. An Application to Consumption Behaviour", **European Economic Review**, 31, 1443-1463.

[208] Steel, M.J. (1987b), **A Bayesian Analysis of Multivariate Exogeneity**, Ph.D. thesis, Université Catholique de Louvain-La-Neuve.

[209] Steel, M.J. and J.F. Richard (1991), "Bayesian Multivariate Exogeneity Analysis: an Application to a UK Money Demand Equation", **Journal of Econometrics**, 49, 239-274.

[210] Stern, R., J. Francis and B. Schumaker (1976), **Price Elasticities in International Trade: An Annotated Bibliography**, MacMillan, London.

[211] Stock, J.H. (1987), "Asymptotic Properties of Least Squares Estimators of Cointegrated Vectors", **Econometrica**, 55, 1035-1056.

[212] Stock, J. and M. Watson (1988), "Testing For Common Trends", **Journal of the American Statistical Association**, 83, 1078-1107.

[213] Stock, J. and M. Watson (1989), "Interpreting The Evidence on Money-Income Causality", **Journal of Econometrics**, 40, 161-181.

[214] Stock, J. and M. Watson (1991), "A simple MLE of Cointegrating Vectors in Higher Order Integrated Systems", NBER Technical Working Paper 83.

[215] Thusby, J. and M. Thursby (1984), "How Reliable are Simple, Single Equation Specifications of Imports Demand ?", **Review of Economics and Statistics**, 66, 120-128.

[216] Todd, R.M. (1990), "Improving Economic Forecasts with Bayesian Vector Autoregression" in: C.W.J. Granger (ed.), **Modelling Economic Series**, Chapter 10, Oxford: Oxford University Press, 214-234.

[217] Urbain, J.-P. (1988), **The Econometric Analysis of Import Demand Functions : an application of cointegration analysis and exogeneity tests**, unpublished master thesis, Université Catholique de Louvain-La-Neuve.

[218] Urbain, J.-P. (1989), "Model Selection Criteria and Granger Causality Tests", **Economics Letters**, 52, 156-162.

[219] Urbain, J.-P. (1990), "Modèles à Correction d'erreur et Fonctions d'importations Agreégées" , **Economie et Prévision, 89/90**, 145-165.

[220] Urbain, J.-P. (1991), "The Behaviour of Some Unit Root Tests in the Presence of Conditional Heteroscedasticity", CREDEL Research Paper 9103, Université de Liège.

[221] Urbain, J.-P. (1992a), "Error Correction Models for Aggregate Imports", in: M. Dagenais and P.A. Muet (eds.), **International Trade Modelling**, Chapter 11, London: Chapman and Hall.

[222] Urbain, J.-P. (1992b), "On Weak Exogeneity in Error Correction Models", **Oxford Bulletin of Economics and Statistics**, 54, 187-207.

[223] Urbain, J.-P. (1992c), "On The Use of Relative Prices in Dynamic Aggregate Import Demand Models", mimeo, University of Limburg.

[224] Urbain, J.-P. (1992d), "Partial versus Full System Analysis of Cointegrated Systems", mimeo, University of Limburg.

[225] White, H. (1980), "A Heteroskedasticity-Consistent Covariance Matrix Estimator and a Direct test for Heteroskedasticity", **Econometrica**, 48, 817-838.

[226] White, H. (1984), **Asymptotic Theory for Econometricians**, Academic Pres, New York.

[227] Wilson, J. and W. Tackacs (1979), "Differential Responses to Prices and Exchange Rates Influences in the Foreign Trade of Selected Industrial Countries", **Review of Economics and Statistics**, 51, 96-104.

[228] Wu, D. (1973), "Alternative Tests of Independence Between Stochastic Regressors and Disturbances", **Econometrica**, 41, 733-750.

[229] Yang, Y.Y. (1991), "Structural Change on the U.S. Import Prices of Manufactures", **Weltwirschaftliches Archiv**, 91, 323-341.

[230] Zellner, A. and F.C. Palm (1974), "Time Series Analysis and Simultaneous Equation Econometric Models", **Journal of Econometrics**, 2, 17-54.

[231] Zellner, A., L. Bauwens and H.K. van Dijk (1988), "Bayesian Specification Analysis and Estimation of Simultaneous Equation Models using Monte Carlo Methods", **Journal of Econometrics**, 38, 39-72.

Author Index

Aldrich, J., 119
Alogoskoufis, G., 38-40
Anderson, T.W., 9
Anderson, G.J., 118
Asseery, A., 114, 118, 121
Banerjee, A., 4, 28, 38, 104
Bärdsen, G., 152
Barndorff-Nielsen, O., 45
Bauwens, L., 85, 88, 89
Bera, A.K., 78, 152, 153
Billinglsey, P., 104
Boswijk, H.P., 2, 3, 4, 5, 28, 36-38, 43, 54, 58, 59, 67, 83, 84, 90, 104, 107-111, 138, 139, 149, 151, 156
Box, G.E.P., 1
Campbell, J.Y., 2, 4, 19, 24
Capuccio, N., 88
Cheng, H.S., 113
Chow, G.C., 34
Clarida, R.H., 114, 117, 120, 121, 122
Clements, M.P., 28, 32, 35, 36
Cuthbertson, K., 12
Davidson, J., 10, 12, 24, 28, 36, 38
Delhausse, B., 24, 133
Dhrymes, P.J., 8
Dickey , D.A., 129-131
Diebold, F.X., 4
Dinter, U., 36
Dolado, J.J., 3, 19, 36, 38, 81, 94, 104
Domowitz, J., 39
Durlauf, S., 24, 104
Engle, R.F., 2, 4, 14, 15, 22, 24, 25, 27, 38, 44, 45, 46, 49, 50, 53, 54, 59, 60, 61, 62, 67, 72, 81, 83, 85, 88, 89, 158
Epstein, R.J., 9
Ericsson, N.R., 12, 15, 36, 38, 49, 94, 119, 158
Escribano, A., 17
Evans, G.B.A., 32
Florens, J.-P., 12, 35, 46
Franses, P.H., 38, 128, 134, 138, 138
Fuller, W.A., 10, 19, 129, 131, 132
Gagnon, J., 114, 116, 117, 118, 120, 123, 150
Galbraith, J.W., 38
Geweke, J., 44, 59, 85
Ghysels, E., 128
Giannini, C., 59
Gilbert, C.L., 25
Ginman, P.G., 122
Goldstein, M., 113, 114, 115, 116, 117, 119, 120, 121, 139
Gonzalo, J., 2
Gourieroux, C., 1, 10, 17, 21, 24,

25, 28, 39, 46, 60, 87
Granger, C.W.J., 2, 4, 5, 10, 17, 22, 23, 38, 43, 53, 54, 67, 72
Gregoire, S., 28, 31
Gregory, A.W., 2
Hall, S., 36, 38, 12
Hansen, B.E., 29, 43, 58
Harvey, A.C., 10
Hausman, J.A., 4, 84, 91, 93
Haynes, S., 117
Hendry, D.F., 3, 5, 10, 12, 14, 15, 29? 32-35, 38, 40, 44-51, 54, 59-62, 64, 81, 83, 85, 89, 104, 119, 120, 123, 137, 142, 143, 158, 164
Holly, A., 45, 46, 64, 81, 88, 89, 91
Hood, W., 9
Horvarth, B., 85
Hunter, J., 54
Husted, S., 114, 117, 120, 123, 150
Hsiao, C., 37
Hylleberg, S., 27, 28, 30, 53, 84, 128
Italianer, A., 142
Jarque, C.M., 78, 152, 153
Jenkins, G.M., 1
Johansen, S., 3, 27-32, 35, 36, 41, 52, 54, 57, 58, 60, 64, 67, 74, 81, 84, 104, 107, 109, 110, 111, 121, 138, 140, 156
Johnston, J.J., 1
Junz, H.B., 120
Juselius, K., 27-32, 35, 36, 67, 74, 138, 140
Keil, M.W., 40
Khan, M., 113, 114, 115, 116, 117, 119, 120, 121, 139
Kholi, U., 117
Kim, Y., 149
Kim, K., 132
King, R.G., 24
Kirchgässner, G., 35
Kiviet, J., 50, 84, 92, 93, 97, 101
Kloek, T., 16
Kollintzas, T.E., 114, 117, 120, 123, 150
Koopmans, T., 9, 45
Krämer, W., 84
Kremers, J.J.M., 3, 29, 36, 38, 94
Lahaye, J.M., 15
Laroque, G., 28, 31
Leamer, E.E., 113
Leipnik, R., 9
Liu, T.C., 10
Loretan, M., 2, 3, 4, 29? 41, 43, 59, 122
Lubrano, M., 48-50, 64, 89, 90, 96, 135
Lucas, R., 11, 119
Lütkepohl, H., 11
McNeilly, C., 149
Magee, S., 117
Malinvaud, E., 9, 35
Marquez, J., 117-119, 149
Mayes, G., 114
Mizon, G., 5, 27, 28, 30, 32, 35, 53, 62, 118, 123, 137, 142, 143, 164
Mouchart, M., 12, 35, 46
Murray, T., 122
Monfort, A., 10, 15, 32, 38, 46, 60, 62, 87
Mosconi, R., 59
Nakamura, A., 91, 93
Nakamura, M., 91, 93
Neale, A.J., 15

Nelson, C.R., 17
Nerlove, M., 4
Newbold, P., 10, 19
Newey, W. K., 131
Nickell, S., 39, 40
Ogaki, M., 2, 31, 138, 143
Oliviera, J., 116
Orsi, R., 88
Ouliaris, S., 29, 131
Orcutt, G., 117, 117
Pagan, A., 12, 39, 120
Palm, F.C., 10, 11, 15, 21, 24, 33, 138
Pfann, G.A., 24, 33
Park, J.Y., 1, 2, 17, 29, 31, 34, 37, 83, 102, 104, 105, 138, 143
Pesaran, H.M., 64, 81, 84
Perron, P., 2, 19, 128, 130, 131, 132, 133, 134
Phillips, P.C.B., 1, 2, 3, 4, 19, 21, 24, 25, 29, 41, 43, 58, 59, 83, 102, 104, 105, 122, 130, 131, 132
Plosser, C., 17
Poirier, D., 5
Rabemanajara, R., 32, 62
Rebelo, S., 24
Reimers, H.E., 2, 138, 139
Richard, J.F., 3, 5, 12, 14, 15, 33, 40, 44-51, 58, 59, 61, 62, 64, 73-80, 83, 85, 87, 89, 90, 96, 111, 119
Rhomberg, R.R., 120
Rubin, H., 9
Saikkonen, P., 34, 43, 58
Sargan, J.D., 24, 38, 39, 120
Sargent, T., 120
Savin, N.E., 21, 110
Sbra, F., 15
Schmidt, P., 132
Schwert, W., 133
Schotman, P., 133
Shiller, R.J., 24
Sims, C., 10, 33
Smith, R., 64, 81, 84
Sneek, J.M., 21
Spanos, A., 5, 14, 15, 17, 33, 51, 64, 134, 137
Steel, M.J., 50-64, 68, 69, 71, 72, 87, 88, 89, 90, 96, 111
Stern, R., 113
Stock, J.H., 24, 26, 29, 43, 121
Stone, J., 117
Tackacs, W., 123
Thursby, J., 120
Thursby, M., 120
Toda, Y., 59
Todd, R.M., 11
Urbain, J.-P., 4, 15, 44, 47, 50, 59, 72, 95, 108, 114, 118, 122, 134, 140, 142, 153, 160
van Dijk, H.K., 85, 88, 89
von Urgern-Stenberg, T., 15
Watson, M., 26, 29, 43, 121
Weiss, A., 2, 25
West, K.D., 131
White, H., 19, 152
Wilson, J., 123
Wu, D., 90
Yang, Y., 149
Yoo, S.B., 4, 27, 29, 81
Zellner, A., 10, 89

Subject Index

ADL model, 15, 56, 62

AR(I)MA processes, 10, 22

ARMAX model, 59

Brownian motion, 104

Chow tests, 146, 146, 147, 158, 159

Cointegration
  Definition, 22
  Cointegrating exogeneity, 54
  Cointegrating rank, 25, 30, 52, 78, 138
  Cointegating vector, 22, 53, 142, 143, 152
  Cointegrated process, 23
  Cointegrated systems, 24-41
  Testing for..., 31, 37, 138, 139, 151

Common trends, 25, 26, 29

Conditional models 34-41, 149
  Conditioning, 12, 14, 35, 45-58
  Partial systems, 34, 123, 149

Consumption function, 68-72

Data generating process (DGP), 12, 13, 45, 51, 93, 94

Dickey-Fuller test, 128
  Sequential ADF tests, 130-132

Encompassing, 33, 144, 145

Error Correction
  Coefficient, 37, 150
  ECMs, 16, 17, 38-40
  Conditional ECMs, 34, 51, 54, 56
  Structural ECMs, 34, 36, 61, 63, 108, 149, 150, 155, 157
  Triangular ECMs, 58

Exogeneity
  Weak exogeneity, 14, 36, 43-81
  Strong exogeneity, 14, 58
  Super exogeneity, 14, 60, 158
  Strict exogeneity, 40
  Testing for..., 67, 71, 83, 89, 107, 153, 154, 156

Factor loading, 30, 52, 67, 78, 141

General to specific, 13, 15

Granger non-causality, 14, 46, 58, 59

Granger representation theorem (GRT), 25-28, 31, 109

Hausman test, 92-106

Imperfect substitutes model, 114-116

Import demand functions
UK import demand, 72
Belgian import demand, 128-164
Dynamic modelling of..., 119

Incomplete simultaneous equation models, 85
Likelihood function, 86, 87
Model representation, 88, 110

Indentification
Conditions for..., 33, 37, 54
Overidentification restrictions, 33, 144, 145

Innovations, 13

Instability, 117, 159, 160

Integrated process, 18

Interim multiplier representation, 28, 30

Invariance principle, 103

Lag length selection, 138, 139

Limited information, 64, 71, 89, 91, 109

Long run equilibrium, 17, 23, 24

Lucas critique, 11, 14, 39, 118

Marginalisation, 12, 35, 134

Maximum eigenvalue test, 31, 139

Money demand, 73-80

Monte Carlo simulation, 93-100

Orthogonality tests, 65, 66, 83, 90, 108-110
Small sample behaviour, 93
Asymptotic distribution, 101

Predeterminedness, 49, 61

Price homogeneity, 122, 123, 140, 152, 160

Random walk, 18

Relative prices, 122, 123, 140, 152, 160

Seasonality, 128-130

Simultaneous equations models, 9, 10, 15, 33, 34, 85, 91, 137, 145, 147, 148

Simultaneity bias, 117

Spurious regression, 19, 24

Stationarity, 17, 20

Structural break, 133, 134, 135

Unit Root Tests, 128-136

VAR models, 10, 11, 13, 26, 30, 53, 54, 57, 78, 138, 142

Variation free, 45, 57

Springer-Verlag and the Environment

We at Springer-Verlag firmly believe that an international science publisher has a special obligation to the environment, and our corporate policies consistently reflect this conviction.

We also expect our business partners – paper mills, printers, packaging manufacturers, etc. – to commit themselves to using environmentally friendly materials and production processes.

The paper in this book is made from low- or no-chlorine pulp and is acid free, in conformance with international standards for paper permanency.

Lecture Notes in Economics and Mathematical Systems

For information about Vols. 1–210
please contact your bookseller or Springer-Verlag

Vol. 211: P. van den Heuvel, The Stability of a Macroeconomic System with Quantity Constraints. VII, 169 pages. 1983.

Vol. 212: R. Sato and T. Nôno, Invariance Principles and the Structure of Technology. V, 94 pages. 1983.

Vol. 213: Aspiration Levels in Bargaining and Economic Decision Making. Proceedings, 1982. Edited by R. Tietz. VIII, 406 pages. 1983.

Vol. 214: M. Faber, H. Niemes und G. Stephan, Entropie, Umweltschutz und Rohstoffverbrauch. IX, 181 Seiten. 1983.

Vol. 215: Semi-Infinite Programming and Applications. Proceedings, 1981. Edited by A.V. Fiacco and K.O. Kortanek. XI, 322 pages. 1983.

Vol. 216: H.H. Müller, Fiscal Policies in a General Equilibrium Model with Persistent Unemployment. VI, 92 pages. 1983.

Vol. 217: Ch. Grootaert, The Relation Between Final Demand and Income Distribution. XIV, 105 pages. 1983.

Vol 218: P. van Loon, A Dynamic Theory of the Firm: Production, Finance and Investment. VII, 191 pages. 1983.

Vol. 219: E. van Damme, Refinements of the Nash Equilibrium Concept. VI. 151 pages. 1983.

Vol. 220: M. Aoki, Notes on Economic Time Series Analysis: System Theoretic Perspectives. IX, 249 pages. 1983.

Vol. 221: S. Nakamura, An Inter-Industry Translog Model of Prices and Technical Change for the West German Economy. XIV, 290 pages. 1984.

Vol. 222: P. Meier, Energy Systems Analysis for Developing Countries. VI, 344 pages. 1984.

Vol. 223: W. Trockel, Market Demand. VIII, 205 pages. 1984.

Vol. 224: M. Kiy, Ein disaggregiertes Prognosesystem fur die Bundesrepublik Deutschland. XVIII, 276 Seiten. 1984.

Vol. 225: T.R. von Ungern-Sternberg, Zur Analyse von Märkten mit unvollständiger Nachfragerinformaton. IX, 125 Seiten. 1984.

Vol. 226: Selected Topics in Operations Research and Mathematical Economics. Proceedings, 1963. Edited by G. Hammer and D. Pallaschke IX, 478 pages. 1984.

Vol. 227: Risk and Capital. Proceedings, 1983. Edited by G. Bamberg and K. Spremann VII, 306 pages. 1984.

Vol. 228: Nonlinear Models of Fluctuating Growth. Proceedings, 1983. Edited by R.M. Goodwin, M. Krüger and A. Vercelli. XVII, 277 pages. 1984.

Vol. 229: Interactive Decision Analysis. Proceedings, 1983. Edited by M. Grauer and A.P. Wierzbicki. VIII, 269 pages. 1984.

Vol. 230: Macro-Economic Planning with Conflicting Goals. Proceedings, 1982. Edited by M. Despontin, P. Nijkamp and J. Spronk. VI, 297 pages. 1984.

Vol. 231: G.F. Newell, The M/M/8 Service System with Ranked Servers in Heavy Traffic. XI, 126 pages. 1984.

Vol. 232: L. Bauwens, Bayesian Full Information Analysis of Simultaneous Equation Models Using Integration by Monte Carlo. VI, 114 pages. 1984.

Vol. 233: G. Wagenhals, The World Copper Market. XI, 190 pages. 1984.

Vol. 234: B.C. Eaves, A Course in Triangulations for Solving Equations with Deformations. III, 302 pages. 1984.

Vol. 235: Stochastic Models in Reliability Theory Proceedings, 1984. Edited by S. Osaki and Y. Hatoyama. VII, 212 pages. 1984.

Vol. 236: G. Gandolfo, P.C. Padoan, A Disequilibrium Model of Real and Financial Accumulation in an Open Economy. VI, 172 pages. 1984.

Vol. 237: Misspecification Analysis. Proceedings, 1983. Edited by T.K. Dijkstra. V, 129 pages. 1984.

Vol. 238: W. Domschke, A. Drexl, Location and Layout Planning. IV, 134 pages. 1985.

Vol. 239: Microeconomic Models of Housing Markets. Edited by K. Stahl. VII, 197 pages. 1985.

Vol. 240: Contributions to Operations Research. Proceedings, 1984. Edited by K. Neumann and D. Pallaschke. V, 190 pages. 1985.

Vol. 241: U. Wittmann, Das Konzept rationaler Preiserwartungen. XI, 310 Seiten. 1985.

Vol. 242: Decision Making with Multiple Objectives. Proceedings, 1984. Edited by Y.Y. Haimes and V. Chankong. XI, 571 pages. 1985.

Vol. 243: Integer Programming and Related Areas. A Classified Bibliography 1981–1984. Edited by R. von Randow. XX, 386 pages. 1985.

Vol. 244: Advances in Equilibrium Theory. Proceedings, 1984. Edited by C.D. Aliprantis, O. Burkinshaw and N.J. Rothman. II, 235 pages. 1985.

Vol. 245: J.E.M. Wilhelm, Arbitrage Theory. VII, 114 pages. 1985.

Vol. 246: P.W. Otter, Dynamic Feature Space Modelling, Filtering and Self-Tuning Control of Stochastic Systems. XIV, 177 pages. 1985.

Vol. 247: Optimization and Discrete Choice in Urban Systems. Proceedings, 1983. Edited by B.G. Hutchinson, P. Nijkamp and M. Batty VI, 371 pages. 1985.

Vol. 248: Pural Rationality and Interactive Decision Processes. Proceedings, 1984. Edited by M. Grauer, M. Thompson and A.P. Wierzbicki. VI, 354 pages. 1985.

Vol. 249: Spatial Price Equilibrium: Advances in Theory, Computation and Application. Proceedings, 1984. Edited by P.T. Harker. VII, 277 pages. 1985.

Vol. 250: M. Roubens, Ph. Vincke, Preference Modelling. VIII, 94 pages. 1985.

Vol. 251: Input-Output Modeling. Proceedings, 1984. Edited by A. Smyshlyaev. VI, 261 pages. 1985.

Vol. 252: A. Birolini, On the Use of Stochastic Processes in Modeling Reliability Problems. VI, 105 pages. 1985.

Vol. 253: C. Withagen, Economic Theory and International Trade in Natural Exhaustible Resources. VI, 172 pages. 1985.

Vol. 254: S. Müller, Arbitrage Pricing of Contingent Claims. VIII, 151 pages. 1985.

Vol. 255: Nondifferentiable Optimization: Motivations and Applications. Proceedings, 1984. Edited by V.F. Demyanov and D. Pallaschke. VI, 350 pages. 1985.

Vol. 256: Convexity and Duality in Optimization. Proceedings, 1984. Edited by J. Ponstein. V, 142 pages. 1985.

Vol. 257: Dynamics of Macrosystems. Proceedings, 1984. Edited by J.-P. Aubin, D. Saari and K. Sigmund. VI, 280 pages. 1985.

Vol. 258: H. Funke, Eine allgemeine Theorie der Polypol- und Oligopolpreisbildung. III, 237 pages. 1985.

Vol. 259: Infinite Programming. Proceedings, 1984. Edited by E.J. Anderson and A.B. Philpott. XIV, 244 pages. 1985.

Vol. 260: H.-J. Kruse, Degeneracy Graphs and the Neighbourhood Problem. VIII, 128 pages. 1986.

Vol. 261: Th.R. Gulledge, Jr., N.K. Womer, The Economics of Made-to-Order Production. VI, 134 pages. 1986.

Vol. 262: H.U. Buhl, A Neo-Classical Theory of Distribution and Wealth. V, 146 pages. 1986.

Vol. 263: M. Schäfer, Resource Extraction and Market Struucture. XI, 154 pages. 1986.

Vol. 264: Models of Economic Dynamics. Proceedings, 1983. Edited by H.F. Sonnenschein. VII, 212 pages. 1986.

Vol. 265: Dynamic Games and Applications in Economics. Edited by T. Basar. IX, 288 pages. 1986.

Vol. 266: Multi-Stage Production Planning and Inventory Control. Edited by S. Axsäter, Ch. Schneeweiss and E. Silver. V, 264 pages.1986.

Vol. 267: R. Bemelmans, The Capacity Aspect of Inventories. IX, 165 pages. 1986.

Vol. 268: V. Firchau, Information Evaluation in Capital Markets. VII, 103 pages. 1986.

Vol. 269: A. Borglin, H. Keiding, Optimality in Infinite Horizon Economies. VI, 180 pages. 1986.

Vol. 270: Technological Change, Employment and Spatial Dynamics. Proceedings, 1985. Edited by P. Nijkamp. VII, 466 pages. 1986.

Vol. 271: C. Hildreth, The Cowles Commission in Chicago, 1939–1955. V, 176 pages. 1986.

Vol. 272: G. Clemenz, Credit Markets with Asymmetric Information. VIII,212 pages. 1986.

Vol. 273: Large-Scale Modelling and Interactive Decision Analysis. Proceedings, 1985. Edited by G. Fandel, M. Grauer, A. Kurzhanski and A.P. Wierzbicki. VII, 363 pages. 1986.

Vol. 274: W.K. Klein Haneveld, Duality in Stochastic Linear and Dynamic Programming. VII, 295 pages. 1986.

Vol. 275: Competition, Instability, and Nonlinear Cycles. Proceedings, 1985. Edited by W. Semmler. XII, 340 pages. 1986.

Vol. 276: M.R. Baye, D.A. Black, Consumer Behavior, Cost of Living Measures, and the Income Tax. VII, 119 pages. 1986.

Vol. 277: Studies in Austrian Capital Theory, Investment and Time. Edited by M. Faber. VI, 317 pages. 1986.

Vol. 278: W.E. Diewert, The Measurement of the Economic Benefits of Infrastructure Services. V, 202 pages. 1986.

Vol. 279: H.-J. Büttler, G. Frei and B. Schips, Estimation of Disequilibrium Modes. VI, 114 pages. 1986.

Vol. 280: H.T. Lau, Combinatorial Heuristic Algorithms with FORTRAN. VII, 126 pages. 1986.

Vol. 281: Ch.-L. Hwang, M.-J. Lin, Group Decision Making under Multiple Criteria. XI, 400 pages. 1987.

Vol. 282: K. Schittkowski, More Test Examples for Nonlinear Programming Codes. V, 261 pages. 1987.

Vol. 283: G. Gabisch, H.-W. Lorenz, Business Cycle Theory. VII, 229 pages. 1987.

Vol. 284: H. Lütkepohl, Forecasting Aggregated Vector ARMA Processes. X, 323 pages. 1987.

Vol. 285: Toward Interactive and Intelligent Decision Support Systems. Volume 1. Proceedings, 1986. Edited by Y. Sawaragi, K. Inoue and H. Nakayama. XII, 445 pages. 1987.

Vol. 286: Toward Interactive and Intelligent Decision Support Systems. Volume 2. Proceedings, 1986. Edited by Y. Sawaragi, K. Inoue and H. Nakayama. XII, 450 pages. 1987.

Vol. 287: Dynamical Systems. Proceedings, 1985. Edited by A.B. Kurzhanski and K. Sigmund. VI, 215 pages. 1987.

Vol. 288: G.D. Rudebusch, The Estimation of Macroeconomic Disequilibrium Models with Regime Classification Information. VII,128 pages. 1987.

Vol. 289: B.R. Meijboom, Planning in Decentralized Firms. X, 168 pages. 1987.

Vol. 290: D.A. Carlson, A. Haurie, Infinite Horizon Optimal Control. XI, 254 pages. 1987.

Vol. 291: N. Takahashi, Design of Adaptive Organizations. VI, 140 pages. 1987.

Vol. 292: I. Tchijov, L. Tomaszewicz (Eds.), Input-Output Modeling. Proceedings, 1985. VI, 195 pages. 1987.

Vol. 293: D. Batten, J. Casti, B. Johansson (Eds.), Economic Evolution and Structural Adjustment. Proceedings, 1985. VI, 382 pages.

Vol. 294: J. Jahn, W. Knabs (Eds.), Recent Advances and Historical Development of Vector Optimization. VII, 405 pages. 1987.

Vol. 295. H. Meister, The Purification Problem for Constrained Games with Incomplete Information. X, 127 pages. 1987.

Vol. 296: A. Börsch-Supan, Econometric Analysis of Discrete Choice. VIII, 211 pages. 1987.

Vol. 297: V. Fedorov, H. Läuter (Eds.), Model-Oriented Data Analysis. Proceedings, 1987. VI, 239 pages. 1988.

Vol. 298: S.H. Chew, Q. Zheng, Integral Global Optimization. VII, 179 pages. 1988.

Vol. 299: K. Marti, Descent Directions and Efficient Solutions in Discretely Distributed Stochastic Programs. XIV, 178 pages. 1988.

Vol. 300: U. Derigs, Programming in Networks and Graphs. XI, 315 pages. 1988.

Vol. 301: J. Kacprzyk, M. Roubens (Eds.), Non-Conventional Preference Relations in Decision Making. VII, 155 pages. 1988.

Vol. 302: H.A. Eiselt, G. Pederzoli (Eds.), Advances in Optimization and Control. Proceedings, 1986. VIII, 372 pages. 1988.

Vol. 303: F.X. Diebold, Empirical Modeling of Exchange Rate Dynamics. VII, 143 pages. 1988.

Vol. 304: A. Kurzhanski, K. Neumann, D. Pallaschke (Eds.), Optimization, Parallel Processing and Applications. Proceedings, 1987. VI, 292 pages. 1988.

Vol. 305: G.-J.C.Th. van Schijndel, Dynamic Firm and Investor Behaviour under Progressive Personal Taxation. X, 215 pages.1988.

Vol. 306: Ch. Klein, A Static Microeconomic Model of Pure Competition. VIII, 139 pages. 1988.

Vol. 307: T.K. Dijkstra (Ed.), On Model Uncertainty and its Statistical Implications. VII, 138 pages. 1988.

Vol. 308: J.R. Daduna, A. Wren (Eds.), Computer-Aided Transit Scheduling. VIII, 339 pages. 1988.

Vol. 309: G. Ricci, K. Velupillai (Eds.), Growth Cycles and Multisectoral Economics: the Goodwin Tradition. III, 126 pages. 1988.

Vol. 310: J. Kacprzyk, M. Fedrizzi (Eds.), Combining Fuzzy Imprecision with Probabilistic Uncertainty in Decision Making. IX, 399 pages. 1988.

Vol. 311: R. Färe, Fundamentals of Production Theory. IX, 163 pages. 1988.

Vol. 312: J. Krishnakumar, Estimation of Simultaneous Equation Models with Error Components Structure. X, 357 pages. 1988.

Vol. 313: W. Jammernegg, Sequential Binary Investment Decisions. VI, 156 pages. 1988.

Vol. 314: R. Tietz, W. Albers, R. Selten (Eds.), Bounded Rational Behavior in Experimental Games and Markets. VI, 368 pages. 1988.

Vol. 315: I. Orishimo, G.J.D. Hewings, P. Nijkamp (Eds), Information Technology: Social and Spatial Perspectives. Proceedings 1986. VI, 268 pages. 1988.

Vol. 316: R.L. Basmann, D.J. Slottje, K. Hayes, J.D. Johnson, D.J. Molina, The Generalized Fechner-Thurstone Direct Utility Function and Some of its Uses. VIII, 159 pages. 1988.

Vol. 317: L. Bianco, A. La Bella (Eds.), Freight Transport Planning and Logistics. Proceedings, 1987. X, 568 pages. 1988.

Vol. 318: T. Doup, Simplicial Algorithms on the Simplotope. VIII, 262 pages. 1988.

Vol. 319: D.T. Luc, Theory of Vector Optimization. VIII, 173 pages. 1989.

Vol. 320: D. van der Wijst, Financial Structure in Small Business. VII, 181 pages. 1989.

Vol. 321: M. Di Matteo, R.M. Goodwin, A. Vercelli (Eds.), Technological and Social Factors in Long Term Fluctuations. Proceedings. IX, 442 pages. 1989.

Vol. 322: T. Kollintzas (Ed.), The Rational Expectations Equilibrium Inventory Model. XI, 269 pages. 1989.

Vol. 323: M.B.M. de Koster, Capacity Oriented Analysis and Design of Production Systems. XII, 245 pages. 1989.

Vol. 324: I.M. Bomze, B.M. Pötscher, Game Theoretical Foundations of Evolutionary Stability. VI, 145 pages. 1989.

Vol. 325: P. Ferri, E. Greenberg, The Labor Market and Business Cycle Theories. X, 183 pages. 1989.

Vol. 326: Ch. Sauer, Alternative Theories of Output, Unemployment, and Inflation in Germany: 1960–1985. XIII, 206 pages. 1989.

Vol. 327: M. Tawada, Production Structure and International Trade. V, 132 pages. 1989.

Vol. 328: W. Güth, B. Kalkofen, Unique Solutions for Strategic Games. VII, 200 pages. 1989.

Vol. 329: G. Tillmann, Equity, Incentives, and Taxation. VI, 132 pages. 1989.

Vol. 330: P.M. Kort, Optimal Dynamic Investment Policies of a Value Maximizing Firm. VII, 185 pages. 1989.

Vol. 331: A. Lewandowski, A.P. Wierzbicki (Eds.), Aspiration Based Decision Support Systems. X, 400 pages. 1989.

Vol. 332: T.R. Gulledge, Jr., L.A. Litteral (Eds.), Cost Analysis Applications of Economics and Operations Research. Proceedings. VII, 422 pages. 1989.

Vol. 333: N. Dellaert, Production to Order. VII, 158 pages. 1989.

Vol. 334: H.-W. Lorenz, Nonlinear Dynamical Economics and Chaotic Motion. XI, 248 pages. 1989.

Vol. 335: A.G. Lockett, G. Islei (Eds.), Improving Decision Making in Organisations. Proceedings. IX, 606 pages. 1989.

Vol. 336: T. Puu, Nonlinear Economic Dynamics. VII, 119 pages. 1989.

Vol. 337: A. Lewandowski, I. Stanchev (Eds.), Methodology and Software for Interactive Decision Support. VIII, 309 pages. 1989.

Vol. 338: J.K. Ho, R.P. Sundarraj, DECOMP: an Implementation of Dantzig-Wolfe Decomposition for Linear Programming. VI, 206 pages.

Vol. 339: J. Terceiro Lomba, Estimation of Dynamic Econometric Models with Errors in Variables. VIII, 116 pages. 1990.

Vol. 340: T. Vasko, R. Ayres, L. Fontvieille (Eds.), Life Cycles and Long Waves. XIV, 293 pages. 1990.

Vol. 341: G.R. Uhlich, Descriptive Theories of Bargaining. IX, 165 pages. 1990.

Vol. 342: K. Okuguchi, F. Szidarovszky, The Theory of Oligopoly with Multi-Product Firms. V, 167 pages. 1990.

Vol. 343: C. Chiarella, The Elements of a Nonlinear Theory of Economic Dynamics. IX, 149 pages. 1990.

Vol. 344: K. Neumann, Stochastic Project Networks. XI, 237 pages. 1990.

Vol. 345: A. Cambini, E. Castagnoli, L. Martein, P Mazzoleni, S. Schaible (Eds.), Generalized Convexity and Fractional Programming with Economic Applications. Proceedings, 1988. VII, 361 pages. 1990.

Vol. 346: R. von Randow (Ed.), Integer Programming and Related Areas. A Classified Bibliography 1984–1987. XIII, 514 pages. 1990.

Vol. 347: D. Ríos Insua, Sensitivity Analysis in Multi-objective Decision Making. XI, 193 pages. 1990.

Vol. 348: H. Störmer, Binary Functions and their Applications. VIII, 151 pages. 1990.

Vol. 349: G.A. Pfann, Dynamic Modelling of Stochastic Demand for Manufacturing Employment. VI, 158 pages. 1990.

Vol. 350: W.-B. Zhang, Economic Dynamics. X, 232 pages. 1990.

Vol. 351: A. Lewandowski, V. Volkovich (Eds.), Multiobjective Problems of Mathematical Programming. Proceedings, 1988. VII, 315 pages. 1991.

Vol. 352: O. van Hilten, Optimal Firm Behaviour in the Context of Technological Progress and a Business Cycle. XII, 229 pages. 1991.

Vol. 353: G. Ricci (Ed.), Decision Processes in Economics. Proceedings, 1989. III, 209 pages 1991.

Vol. 354: M. Ivaldi, A Structural Analysis of Expectation Formation. XII, 230 pages. 1991.

Vol. 355: M. Salomon. Deterministic Lotsizing Models for Production Planning. VII, 158 pages. 1991.

Vol. 356: P. Korhonen, A. Lewandowski, J . Wallenius (Eds.), Multiple Criteria Decision Support. Proceedings, 1989. XII, 393 pages. 1991.

Vol. 357: P. Zörnig, Degeneracy Graphs and Simplex Cycling. XV, 194 pages. 1991.

Vol. 358: P. Knottnerus, Linear Models with Correlated Disturbances. VIII, 196 pages. 1991.

Vol. 359: E. de Jong, Exchange Rate Determination and Optimal Economic Policy Under Various Exchange Rate Regimes. VII, 270 pages. 1991.

Vol. 360: P. Stalder, Regime Translations, Spillovers and Buffer Stocks. VI, 193 pages . 1991.

Vol. 361: C. F. Daganzo, Logistics Systems Analysis. X, 321 pages. 1991.

Vol. 362: F. Gehrels, Essays In Macroeconomics of an Open Economy. VII, 183 pages. 1991.

Vol. 363: C. Puppe, Distorted Probabilities and Choice under Risk. VIII, 100 pages . 1991

Vol. 364: B. Horvath, Are Policy Variables Exogenous? XII, 162 pages. 1991.

Vol. 365: G. A. Heuer, U. Leopold-Wildburger. Balanced Silverman Games on General Discrete Sets. V, 140 pages. 1991.

Vol. 366: J. Gruber (Ed.), Econometric Decision Models. Proceedings, 1989. VIII, 636 pages. 1991.

Vol. 367: M. Grauer, D. B. Pressmar (Eds.), Parallel Computing and Mathematical Optimization. Proceedings. V, 208 pages. 1991.

Vol. 368: M. Fedrizzi, J. Kacprzyk, M. Roubens (Eds.), Interactive Fuzzy Optimization. VII, 216 pages. 1991.

Vol. 369: R. Koblo, The Visible Hand. VIII, 131 pages.1991.

Vol. 370: M. J. Beckmann, M. N. Gopalan, R. Subramanian (Eds.), Stochastic Processes and their Applications. Proceedings, 1990. XLI, 292 pages. 1991.

Vol. 371: A. Schmutzler, Flexibility and Adjustment to Information in Sequential Decision Problems. VIII, 198 pages. 1991.

Vol. 372: J. Esteban, The Social Viability of Money. X, 202 pages. 1991.

Vol. 373: A. Billot, Economic Theory of Fuzzy Equilibria. XIII, 164 pages. 1992.

Vol. 374: G. Pflug, U. Dieter (Eds.), Simulation and Optimization. Proceedings, 1990. X, 162 pages. 1992.

Vol. 375: S.-J. Chen, Ch.-L. Hwang, Fuzzy Multiple Attribute Decision Making. XII, 536 pages. 1992.

Vol. 376: K.-H. Jöckel, G. Rothe, W. Sendler (Eds.), Bootstrapping and Related Techniques. Proceedings, 1990. VIII, 247 pages. 1992.

Vol. 377: A. Villar, Operator Theorems with Applications to Distributive Problems and Equilibrium Models. XVI, 160 pages. 1992.

Vol. 378: W. Krabs, J. Zowe (Eds.), Modern Methods of Optimization. Proceedings, 1990. VIII, 348 pages. 1992.

Vol. 379: K. Marti (Ed.), Stochastic Optimization. Proceedings, 1990. VII, 182 pages. 1992.

Vol. 380: J. Odelstad, Invariance and Structural Dependence. XII, 245 pages. 1992.

Vol. 381: C. Giannini, Topics in Structural VAR Econometrics. XI, 131 pages. 1992.

Vol. 382: W. Oettli, D. Pallaschke (Eds.), Advances in Optimization. Proceedings, 1991. X, 527 pages. 1992.

Vol. 383: J. Vartiainen, Capital Accumulation in a Corporatist Economy. VII, 177 pages. 1992.

Vol. 384: A. Martina, Lectures on the Economic Theory of Taxation. XII, 313 pages. 1992.

Vol. 385: J. Gardeazabal, M. Regúlez, The Monetary Model of Exchange Rates and Cointegration. X, 194 pages. 1992.

Vol. 386: M. Desrochers, J.-M. Rousseau (Eds.), Computer-Aided Transit Scheduling. Proceedings, 1990. XIII, 432 pages. 1992.

Vol. 387: W. Gaertner, M. Klemisch-Ahlert, Social Choice and Bargaining Perspectives on Distributive Justice. VIII, 131 pages. 1992.

Vol. 388: D. Bartmann, M. J. Beckmann, Inventory Control. XV, 252 pages. 1992.

Vol. 389: B. Dutta, D. Mookherjee, T. Parthasarathy, T. Raghavan, D. Ray, S. Tijs (Eds.), Game Theory and Economic Applications. Proceedings, 1990. ??, ?? pages. 1992.

Vol. 390: G. Sorger, Minimum Impatience Theorem for Recursive Economic Models. X, 162 pages. 1992.

Vol. 391: C. Keser, Experimental Duopoly Markets with Demand Inertia. X, 150 pages. 1992.

Vol. 392: K. Frauendorfer, Stochastic Two-Stage Programming. VIII, 228 pages. 1992.

Vol. 393: B. Lucke, Price Stabilization on World Agricultural Markets. XI, 274 pages. 1992.

Vol. 394: Y.-J. Lai, C.-L. Hwang, Fuzzy Mathematical Programming. XIII, 301 pages. 1992.

Vol. 395: G. Haag, U. Mueller, K. G. Troitzsch (Eds.), Economic Evolution and Demographic Change. XVI, 409 pages. 1992.

Vol. 396: R. V. V. Vidal (Ed.), Applied Simulated Annealing. VIII, 358 pages. 1992.

Vol. 397: J. Wessels, A. P. Wierzbicki (Eds.), User-Oriented Methodology and Techniques of Decision Analysis and Support. Proceedings, 1991. XII, 295 pages. 1993.

Vol. 398: J.-P. Urbain, Exogeneity in Error Correction Models. XI, 189 pages. 1993.

GPSR Compliance

The European Union's (EU) General Product Safety Regulation (GPSR) is a set of rules that requires consumer products to be safe and our obligations to ensure this.

If you have any concerns about our products, you can contact us on ProductSafety@springernature.com

In case Publisher is established outside the EU, the EU authorized representative is:

Springer Nature Customer Service Center GmbH
Europaplatz 3
69115 Heidelberg, Germany

Batch number: 10844829

Printed by Printforce, the Netherlands